Animal Anatomy

Animal Anatomy

Edited by
Reuben Tidwell

Larsen & Keller
www.larsen-keller.com

Animal Anatomy
Edited by Reuben Tidwell
ISBN: 978-1-63549-759-5 (Hardback)

Published by Larsen and Keller Education,
5 Penn Plaza,
19th Floor,
New York, NY 10001, USA

Cataloging-in-Publication Data

Animal anatomy / edited by Reuben Tidwell.
     p. cm.
Includes bibliographical references and index.
ISBN 978-1-63549-759-5
1. Anatomy. 2. Animals. 3. Physiology. 4. Veterinary anatomy. I. Tidwell, Reuben.
QL805 .A55 2018
571.3--dc23

For more information regarding Larsen and Keller Education and its products, please visit the publisher's website www.larsen-keller.com

Table of Contents

Permissions

Index

Preface

As a part of life sciences, anatomy refers to the study of structure of living organisms and their body parts. It uses the elements of evolutionary biology, embryology, phylogeny, comparative anatomy, etc. to study, analyze and understand the structure of living beings. Animal anatomy refers to the study of structure of animals. It includes understanding animal tissues, vertebrate anatomy, arthropod anatomy, etc. This book is a compilation of chapters that discuss the most vital concepts in the field of animal anatomy. It outlines the processes and applications of this field in detail. This textbook is meant for students who are looking for an elaborate reference text on animal anatomy.

A foreword of all chapters of the book is provided below:

Chapter 1 - Animal anatomy studies the structure of animals. It can be categorized into fish anatomy, amphibian anatomy, reptilian anatomy, avian anatomy, etc. The chapter on animal anatomy offers an insightful focus, keeping in mind the complex subject matter; **Chapter 2** - Ameloblasts are the cells found during the development of teeth. Enteroendocrine cells are found in the gastrointestinal tract and the pancreas and help with the endocrine function. Cnidocyte, choanocyte, parietal cell, muscle tissue, pericyte and epithelium are the other types of cells and tissues explained in this section. This chapter provides a plethora of interdisciplinary topics for better comprehension of animal anatomy; **Chapter 3** - Lateral lines are found in aquatic animals. It helps them in sensing the movement and vibration which surrounds them in water. The other types of sensory organs and nervous systems explained are cercus, crista acustica, forked tongue, Mauthner cell, medullary command nucleus, etc. The topics elaborated in this chapter will help in gaining a better perspective about the subject matter; **Chapter 4** - Bird anatomy is the study of the internal and external parts of a bird. The subject studies all the systems related to birds. Some examples of these systems are skeletal system, muscular system, respiratory system, digestive system and circulatory system. The topics discussed in the chapter are of great importance to broaden the existing knowledge on bird anatomy; **Chapter 5** - Animal locomotion is the term used for the movement of animals. Some of the methods used for movement by animals are running, jumping, flying, hopping and swimming. Terrestrial locomotion, arboreal locomotion and aquatic locomotion are also discussed. This chapter is an overview of the subject matter incorporating all the major aspects of animal locomotion.

At the end, I would like to thank all the people associated with this book devoting their precious time and providing their valuable contributions to this book. I would also like to express my gratitude to my fellow colleagues who encouraged me throughout the process.

Editor

An Introduction to Animal and their Anatomy

Animal anatomy studies the structure of animals. It can be categorized into fish anatomy, amphibian anatomy, reptilian anatomy, avian anatomy, etc. The chapter on animal anatomy offers an insightful focus, keeping in mind the complex subject matter.

Animal

Animals are multicellular, eukaryotic organisms of the kingdom Animalia (also called Metazoa). The animal kingdom emerged as a clade within Apoikozoa as the sister group to the choanoflagellates. Animals are motile, meaning they can move spontaneously and independently at some point in their lives. Their body plan eventually becomes fixed as they develop, although some undergo a process of metamorphosis later in their lives. All animals are heterotrophs: they must ingest other organisms or their products for sustenance.

Most known animal phyla appeared in the fossil record as marine species during the Cambrian explosion, about 542 million years ago. Animals can be divided broadly into vertebrates and invertebrates. Vertebrates have a backbone or spine (vertebral column), and amount to less than five percent of all described animal species. They include fish, amphibians, reptiles, birds and mammals. The remaining animals are the invertebrates, which lack a backbone. These include molluscs (clams, oysters, octopuses, squid, snails); arthropods (millipedes, centipedes, insects, spiders, scorpions, crabs, lobsters, shrimp); annelids (earthworms, leeches), nematodes (filarial worms, hookworms), flatworms (tapeworms, liver flukes), cnidarians (jellyfish, sea anemones, corals), ctenophores (comb jellies), and sponges. The study of animals is called zoology.

Etymology

The word "animal" comes from the Latin *animalis*, meaning *having breath*, *having soul* or *living being*. In everyday non-scientific usage the word excludes humans – that is, animal is often used to refer only to non-human members of the kingdom Animalia; often, only closer relatives of humans such as mammals and other vertebrates, are meant. The biological definition of the word refers to all members of the kingdom Animalia, encompassing creatures as diverse as sponges, jellyfish, insects, and humans.

History of Classification

Aristotle divided the living world between animals and plants, and this was followed by Carl Linnaeus, in the first hierarchical classification. In Linnaeus's original scheme, the animals were one of three kingdoms, divided into the classes of Vermes, Insecta, Pisces, Amphibia, Aves, and

Mammalia. Since then the last four have all been subsumed into a single phylum, the Chordata, whereas the various other forms have been separated out.

Carl Linnaeus, an animal himself, is known as the father of modern taxonomy

In 1874, Ernst Haeckel divided the animal kingdom into two subkingdoms: Metazoa (multicellular animals) and Protozoa (single-celled animals). The protozoa were later moved to the kingdom Protista, leaving only the metazoa. Thus Metazoa is now considered a synonym of Animalia.

Characteristics

Animals have several characteristics that set them apart from other living things. Animals are eukaryotic and multicellular, which separates them from bacteria and most protists. They are heterotrophic, generally digesting food in an internal chamber, which separates them from plants and algae. They are also distinguished from plants, algae, and fungi by lacking rigid cell walls. All animals are motile, if only at certain life stages. In most animals, embryos pass through a blastula stage, which is a characteristic exclusive to animals.

Structure

With a few exceptions, most notably the sponges (Phylum Porifera) and Placozoa, animals have bodies differentiated into separate tissues. These include muscles, which are able to contract and control locomotion, and nerve tissues, which send and process signals. Typically, there is also an internal digestive chamber, with one or two openings. Animals with this sort of organization are called metazoans, or eumetazoans when the former is used for animals in general.

All animals have eukaryotic cells, surrounded by a characteristic extracellular matrix composed of collagen and elastic glycoproteins. This may be calcified to form structures like shells, bones, and spicules. During development, it forms a relatively flexible framework upon which cells can move about and be reorganized, making complex structures possible. In contrast, other multicellular organisms, like plants and fungi, have cells held in place by cell walls, and so develop by progressive growth. Also, unique to animal cells are the following intercellular junctions: tight junctions, gap junctions, and desmosomes.

Reproduction and Development

Nearly all animals undergo some form of sexual reproduction. They produce haploid gametes by meiosis. The smaller, motile gametes are spermatozoa and the larger, non-motile gametes are ova. These fuse to form zygotes, which develop into new individuals.

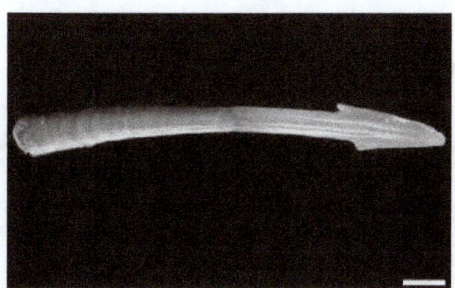

Some species of land snails use love darts as a form of sexual selection

Many animals are also capable of asexual reproduction. This may take place through parthenogenesis, where fertile eggs are produced without mating, budding, or fragmentation.

A zygote initially develops into a hollow sphere, called a blastula, which undergoes rearrangement and differentiation. In sponges, blastula larvae swim to a new location and develop into a new sponge. In most other groups, the blastula undergoes more complicated rearrangement. It first invaginates to form a gastrula with a digestive chamber, and two separate germ layers—an external ectoderm and an internal endoderm. In most cases, a mesoderm also develops between them. These germ layers then differentiate to form tissues and organs.

Inbreeding Avoidance

In Gombe Stream National Park, male chimpanzees remain in their natal community while females disperse to other groups.

During sexual reproduction, mating with a close relative (inbreeding) generally leads to inbreeding depression. For instance, inbreeding was found to increase juvenile mortality in 11 small animal species. Inbreeding depression is considered to be largely due to expression of deleterious recessive mutations. Mating with unrelated or distantly related members of the same species is generally thought to provide the advantage of masking deleterious recessive mutations in progeny. Animals have evolved numerous diverse mechanisms for avoiding close inbreeding and promoting outcrossing.

As indicated in the image of chimpanzees, they have adopted dispersal as a way to separate close relatives and prevent inbreeding. Their dispersal route is known as natal dispersal, whereby individuals move away from the area of birth.

DNA analysis has shown that 60% of offspring in splendid fairywrens nests were sired through extra-pair copulations, rather than from resident males.

In various species, such as the splendid fairywren, females benefit by mating with multiple males, thus producing more offspring of higher genetic quality. Females that are pair bonded to a male of poor genetic quality, as is the case in inbreeding, are more likely to engage in extra-pair copulations in order to improve their reproductive success and the survivability of their offspring.

Food and Energy Sourcing

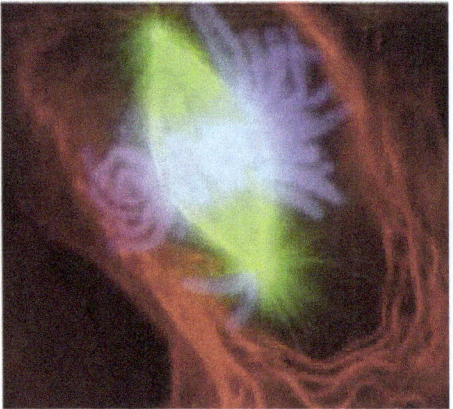

A newt lung cell stained with fluorescent dyes undergoing the early anaphase stage of mitosis.

All animals are heterotrophs, meaning that they feed directly or indirectly on other living things. They are often further subdivided into groups such as carnivores, herbivores, omnivores, and parasites.

Predation is a biological interaction where a predator (a heterotroph that is hunting) feeds on its prey (the organism that is attacked). Predators may or may not kill their prey prior to feeding on them, but the act of predation almost always results in the death of the prey. The other main category of consumption is detritivory, the consumption of dead organic matter. It can at times be difficult to separate the two feeding behaviours, for example, where parasitic species prey on a host organism and then lay their eggs on it for their offspring to feed on its decaying corpse. Selective pressures imposed on one another has led to an evolutionary arms race between prey and predator, resulting in various antipredator adaptations.

Most animals indirectly use the energy of sunlight by eating plants or plant-eating animals. Most plants use light to convert inorganic molecules in their environment into carbohydrates, fats, proteins and other biomolecules, characteristically containing reduced carbon in the form of carbon-hydrogen bonds. Starting with carbon dioxide (CO_2) and water (H_2O), photosynthesis converts the energy of sunlight into chemical energy in the form of simple sugars (e.g., glucose), with the release of molecular oxygen. These sugars are then used as the building blocks for plant growth, including the production of other biomolecules. When an animal eats plants (or eats other animals which have eaten plants), the reduced carbon compounds in the food become a source of energy and building materials for the animal. They are either used directly to help the animal grow, or broken down, releasing stored solar energy, and giving the animal the energy required for motion.

Animals living close to hydrothermal vents and cold seeps on the ocean floor are not dependent on the energy of sunlight. Instead chemosynthetic archaea and bacteria form the base of the food chain.

Origin and Fossil Record

Dunkleosteus was a 10-metre-long (33 ft) prehistoric fish

Animals are generally considered to have emerged within flagellated eukaryota. Their closest known living relatives are the choanoflagellates, collared flagellates that have a morphology similar to the choanocytes of certain sponges. Molecular studies place animals in a supergroup called the opisthokonts, which also include the choanoflagellates, fungi and a few small parasitic protists. The name comes from the posterior location of the flagellum in motile cells, such as most animal spermatozoa, whereas other eukaryotes tend to have anterior flagella.

The first fossils that might represent animals appear in the Trezona Formation at Trezona Bore, West Central Flinders, South Australia. These fossils are interpreted as being early sponges. They were found in 665-million-year-old rock.

The next oldest possible animal fossils are found towards the end of the Precambrian, around 610 million years ago, and are known as the Ediacaran or Vendian biota. These are difficult to relate to later fossils, however. Some may represent precursors of modern phyla, but they may be separate groups, and it is possible they are not really animals at all.

Aside from them, most known animal phyla make a more or less simultaneous appearance during the Cambrian period, about 542 million years ago. It is still disputed whether this event, called the

Cambrian explosion, is due to a rapid divergence between different groups or due to a change in conditions that made fossilization possible.

Some palaeontologists suggest that animals appeared much earlier than the Cambrian explosion, possibly as early as 1 billion years ago. Trace fossils such as tracks and burrows found in the Tonian period indicate the presence of triploblastic worms, like metazoans, roughly as large (about 5 mm wide) and complex as earthworms. During the beginning of the Tonian period around 1 billion years ago, there was a decrease in Stromatolite diversity, which may indicate the appearance of grazing animals, since stromatolite diversity increased when grazing animals became extinct at the End Permian and End Ordovician extinction events, and decreased shortly after the grazer populations recovered. However the discovery that tracks very similar to these early trace fossils are produced today by the giant single-celled protist *Gromia sphaerica* casts doubt on their interpretation as evidence of early animal evolution.

Groups of Animals

Traditional morphological and modern molecular phylogenetic analysis have both recognized a major evolutionary transition from "non-bilaterian" animals, which are those lacking a bilaterally symmetric body plan (Porifera, Ctenophora, Cnidaria and Placozoa), to "bilaterian" animals (Bilateria) whose body plans display bilateral symmetry. The latter are further classified based on a major division between Deuterostomes and Protostomes. The relationships among non-bilaterian animals are disputed, but all bilaterian animals are thought to form a monophyletic group. Current understanding of the relationships among the major groups of animals is summarized by the following cladogram:

Non-bilaterian Animals: Porifera, Placozoa, Ctenophora, Cnidaria

Orange elephant ear sponge, *Agelas clathrodes*, in foreground. Two corals in the background: a sea fan, *Iciligorgia schrammi*, and a sea rod, *Plexaurella nutans*.

Several animal phyla are recognized for their lack of bilateral symmetry, and are thought to have diverged from other animals early in evolution. Among these, the sponges (Porifera) were long thought to have diverged first, representing the oldest animal phylum. They lack the complex organization found in most other phyla. Their cells are differentiated, but in most cases not organized into distinct tissues. Sponges typically feed by drawing in water through pores. However, a series of phylogenomic studies from 2008-2015 have found support for Ctenophora, or comb jellies, as the basal lineage of animals. This result has been controversial, since it would imply that sponges may not be so primitive, but may instead be secondarily simplified. Other researchers have argued that the placement of Ctenophora as the earliest-diverging animal phylum is a statistical anomaly caused by the high rate of evolution in ctenophore genomes.

Among the other phyla, the Ctenophora and the Cnidaria, which includes sea anemones, corals, and jellyfish, are radially symmetric and have digestive chambers with a single opening, which serves as both the mouth and the anus. Both have distinct tissues, but they are not organized into organs. There are only two main germ layers, the ectoderm and endoderm, with only scattered cells between them. As such, these animals are sometimes called diploblastic. The tiny placozoans are similar, but they do not have a permanent digestive chamber.

The Myxozoa, microscopic parasites that were originally considered Protozoa, are now believed to have evolved within Cnidaria.

Bilaterian Animals

The remaining animals form a monophyletic group called the Bilateria. For the most part, they are bilaterally symmetric, and often have a specialized head with feeding and sensory organs. The body is triploblastic, i.e. all three germ layers are well-developed, and tissues form distinct organs. The digestive chamber has two openings, a mouth and an anus, and there is also an internal body cavity called a coelom or pseudocoelom. There are exceptions to each of these characteristics, however—for instance adult echinoderms are radially symmetric, and certain parasitic worms have extremely simplified body structures.

Genetic studies have considerably changed our understanding of the relationships within the Bilateria. Most appear to belong to two major lineages: the deuterostomes and the protostomes, the latter of which includes the Ecdysozoa, and Lophotrochozoa. The Chaetognatha or arrow worms have been traditionally classified as deuterostomes, though recent molecular studies have identified this group as a basal protostome lineage.

In addition, there are a few small groups of bilaterians with relatively cryptic morphology whose relationships with other animals are not well-established. For example, recent molecular studies have identified Acoelomorpha and *Xenoturbella* as comprising a monophyletic group, but studies disagree as to whether this group evolved from within deuterostomes, or whether it represents the sister group to all other bilaterian animals (Nephrozoa). Other groups of uncertain affinity include the Rhombozoa and Orthonectida. One phyla, the Monoblastozoa, was described by a scientist in 1892, but so far there have been no evidence of its existence.

Deuterostomes and Protostomes

Deuterostomes differ from protostomes in several ways. Animals from both groups possess a complete digestive tract. However, in protostomes, the first opening of the gut to appear in embryological development (the archenteron) develops into the mouth, with the anus forming secondarily. In deuterostomes the anus forms first, with the mouth developing secondarily. In most protostomes, cells simply fill in the interior of the gastrula to form the mesoderm, called schizocoelous development, but in deuterostomes, it forms through invagination of the endoderm, called enterocoelic pouching. Deuterostome embryos undergo radial cleavage during cell division, while protostomes undergo spiral cleavage.

Superb fairy-wren, *Malurus cyaneus*

All this suggests the deuterostomes and protostomes are separate, monophyletic lineages. The main phyla of deuterostomes are the Echinodermata and Chordata. The former are radially symmetric and exclusively marine, such as starfish, sea urchins, and sea cucumbers. The latter are dominated by the vertebrates, animals with backbones. These include fish, amphibians, reptiles, birds, and mammals.

In addition to these, the deuterostomes also include the Hemichordata, or acorn worms, which are thought to be closely related to Echinodermata forming a group known as Ambulacraria. Although they are not especially prominent today, the important fossil graptolites may belong to this group.

Ecdysozoa

Yellow-winged darter, *Sympetrum flaveolum*

The Ecdysozoa are protostomes, named after the common trait of growth by moulting or ecdysis. The largest animal phylum belongs here, the Arthropoda, including insects, spiders, crabs, and their kin. All these organisms have a body divided into repeating segments, typically with paired appendages. Two smaller phyla, the Onychophora and Tardigrada, are close relatives of the arthropods and share these traits. The ecdysozoans also include the Nematoda or roundworms, perhaps the second largest animal phylum. Roundworms are typically microscopic, and occur in nearly every environment where there is water. A number are important parasites. Smaller phyla related to them are the Nematomorpha or horsehair worms, and the Kinorhyncha, Priapulida, and Loricifera. These groups have a reduced coelom, called a pseudocoelom.

Roman snail, *Helix pomatia*

Lophotrochozoa

The Lophotrochozoa, evolved within Protostomia, include two of the most successful animal phyla, the Mollusca and Annelida. The former, which is the second-largest animal phylum by number of described species, includes animals such as snails, clams, and squids, and the latter comprises the segmented worms, such as earthworms and leeches. These two groups have long been considered close relatives because of the common presence of trochophore larvae, but the annelids were considered closer to the arthropods because they are both segmented. Now, this is generally considered convergent evolution, owing to many morphological and genetic differences between the two phyla. Lophotrochozoa also includes the Nemertea or ribbon worms, the Sipuncula, and several phyla that have a ring of ciliated tentacles around the mouth, called a lophophore. These were traditionally grouped together as the lophophorates. but it now appears that the lophophorate group may be paraphyletic, with some closer to the nemerteans and some to the molluscs and annelids. They include the Brachiopoda or lamp shells, which are prominent in the fossil record, the Entoprocta, the Phoronida, and possibly the Bryozoa or moss animals.

The Platyzoa include the phylum Platyhelminthes, the flatworms. These were originally considered some of the most primitive Bilateria, but it now appears they developed from more complex ancestors. A number of parasites are included in this group, such as the flukes and tapeworms. Flatworms are acoelomates, lacking a body cavity, as are their closest relatives, the microscopic Gastrotricha. The other platyzoan phyla are mostly microscopic and pseudocoelomate. The most prominent are the Rotifera or rotifers, which are common in aqueous environments. They also include the Acanthocephala or spiny-headed worms, the Gnathostomulida, Micrognathozoa, and possibly the Cycliophora. These groups share the presence of complex jaws, from which they are called the Gnathifera.

A relationship between the Brachiopoda and Nemertea has been suggested by molecular data. A second study has also suggested this relationship. This latter study also suggested that Annelida and Mollusca may be sister clades. Another study has suggested that Annelida and Mollusca are sister clades. This clade has been termed the Neotrochozoa.

Number of Extant Species

Animals can be divided into two broad groups: vertebrates (animals with a backbone) and invertebrates (animals without a backbone). Half of all described vertebrate species are fishes and three-quarters of all described invertebrate species are insects. The following table lists the number of described extant species for each major animal subgroup as estimated for the IUCN Red List of Threatened Species, *2014*.

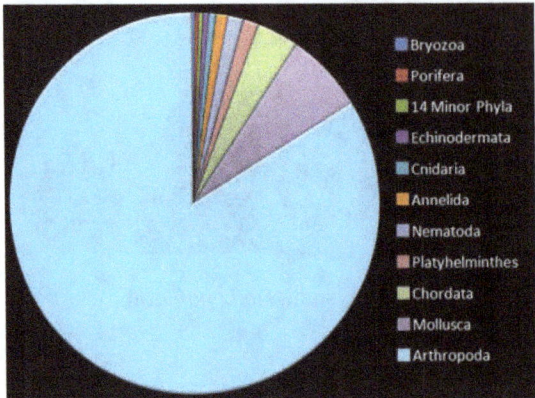

The relative number of species contributed to the total by each phylum of animals

Group	Image	Subgroup	Estimated number of described species
Vertebrates		Fishes	32,900
		Amphibians	7,302
		Reptiles	10,038
		Birds	10,425
		Mammals	5,513
Total vertebrate species: 66,178			

Invertebrates		Insects	1,000,000
		Molluscs	85,000
		Crustaceans	47,000
Invertebrates		Corals	2,000
		Arachnids	102,248
		Velvet worms	165
		Horseshoe crabs	4
		Others	68,658
Total invertebrate species: 1,305,075			
Total for all animal species: 1,371,253			

Over 95% of the described animal species in the world are invertebrates.

Model Organisms

Because of the great diversity found in animals, it is more economical for scientists to study a small number of chosen species so that connections can be drawn from their work and conclusions extrapolated about how animals function in general. Because they are easy to keep and breed, the fruit fly *Drosophila melanogaster* and the nematode *Caenorhabditis elegans* have long been the most intensively studied metazoan model organisms, and were among the first life-forms to be genetically sequenced. This was facilitated by the severely reduced state of their genomes, but as many genes, introns, and linkages lost, these ecdysozoans can teach us little about the origins of animals in general. The extent of this type of evolution within the superphylum will be revealed by the crustacean, annelid, and molluscan genome projects currently in progress. Analysis of the starlet sea anemone genome has emphasized the importance of sponges, placozoans, and choanoflagellates, also being sequenced, in explaining the arrival of 1500 ancestral genes unique to the Eumetazoa.

An analysis of the homoscleromorph sponge *Oscarella carmela* also suggests that the last common ancestor of sponges and the eumetazoan animals was more complex than previously assumed.

Other model organisms belonging to the animal kingdom include the house mouse (*Mus musculus*), laboratory rat (*Rattus norvegicus*) and zebrafish (*Danio rerio*).

Animal Anatomy

Anatomy is the branch of biology concerned with the study of the structure of organisms and their parts. Anatomy, branch of natural science dealing with the structural organization of living things. It is an old science, having its beginnings in prehistoric times. Anatomy is inherently tied to embryology, comparative anatomy, evolutionary biology, and phylogeny, as these are the processes by which anatomy is generated over immediate (embryology) and long (evolution) timescales. Human anatomy is one of the basic essential sciences of medicine.

The discipline of anatomy is divided into macroscopic and microscopic anatomy. Macroscopic anatomy, or gross anatomy, is the examination of an animal's body parts using unaided eyesight. Gross anatomy also includes the branch of superficial anatomy. Microscopic anatomy involves the use of optical instruments in the study of the tissues of various structures, known as histology, and also in the study of cells.

The history of anatomy is characterized by a progressive understanding of the functions of the organs and structures of the human body. Methods have also improved dramatically, advancing from the examination of animals by dissection of carcasses and cadavers (corpses) to 20th century medical imaging techniques including X-ray, ultrasound, and magnetic resonance imaging.

Anatomy and physiology, which study (respectively) the structure and function of organisms and their parts, make a natural pair of related disciplines, and they are often studied together.

Definition

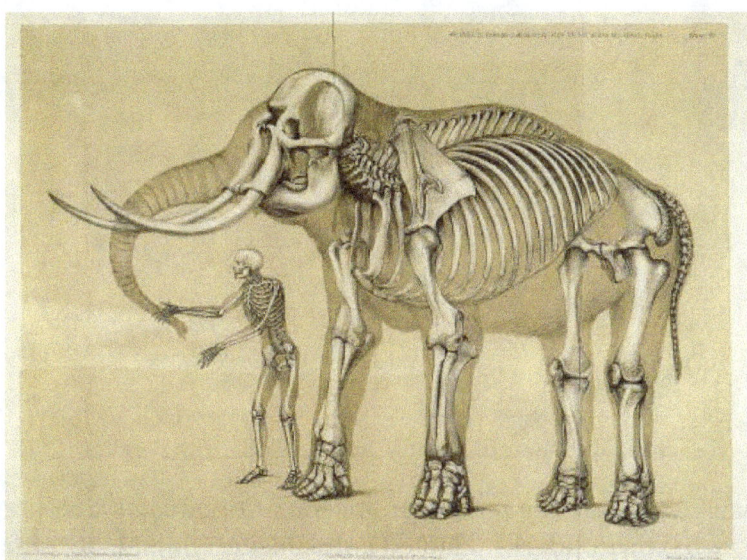

Human compared to elephant frame. Benjamin Waterhouse Hawkins, 1860

Derived from the Greek *anatomē* "dissection", anatomy is the scientific study of the structure of organisms including their systems, organs and tissues. It includes the appearance and position of the various parts, the materials from which they are composed, their locations and their relationships with other parts. Anatomy is quite distinct from physiology and biochemistry, which deal respectively with the functions of those parts and the chemical processes involved. For example, an anatomist is concerned with the shape, size, position, structure, blood supply and innervation of an organ such as the liver; while a physiologist is interested in the production of bile, the role of the liver in nutrition and the regulation of bodily functions.

The discipline of anatomy can be subdivided into a number of branches including gross or macroscopic anatomy and microscopic anatomy. Gross anatomy is the study of structures large enough to be seen with the naked eye, and also includes superficial anatomy or surface anatomy, the study by sight of the external body features. Microscopic anatomy is the study of structures on a microscopic scale, including histology (the study of tissues), and embryology (the study of an organism in its immature condition).

Anatomy can be studied using both invasive and non-invasive methods with the goal of obtaining information about the structure and organization of organs and systems. Methods used include dissection, in which a body is opened and its organs studied, and endoscopy, in which a video camera-equipped instrument is inserted through a small incision in the body wall and used to explore the internal organs and other structures. Angiography using X-rays or magnetic resonance angiography are methods to visualize blood vessels.

The term "anatomy" is commonly taken to refer to human anatomy. However, substantially the same structures and tissues are found throughout the rest of the animal kingdom and the term also includes the anatomy of other animals. The term *zootomy* is also sometimes used to specifically refer to animals. The structure and tissues of plants are of a dissimilar nature and they are studied in plant anatomy.

Animal Tissues

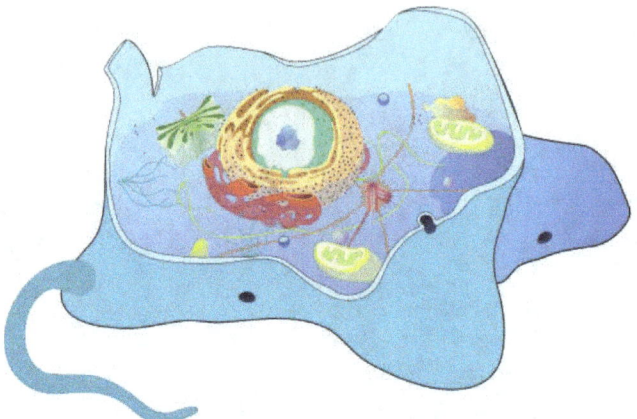

Stylized cutaway diagram of an animal cell (with flagella)

The kingdom Animalia or metazoa, contains multicellular organisms that are heterotrophic and motile (although some have secondarily adopted a sessile lifestyle). Most animals have bodies differentiated into separate tissues and these animals are also known as eumetazoans. They have an internal digestive chamber, with one or two openings; the gametes are produced in multicellular sex organs, and the zygotes include a blastula stage in their embryonic development. Metazoans do not include the sponges, which have undifferentiated cells.

Unlike plant cells, animal cells have neither a cell wall nor chloroplasts. Vacuoles, when present, are more in number and much smaller than those in the plant cell. The body tissues are composed of numerous types of cell, including those found in muscles, nerves and skin. Each typically has a cell membrane formed of phospholipids, cytoplasm and a nucleus. All of the different cells of an animal are derived from the embryonic germ layers. Those simpler invertebrates which are formed from two germ layers of ectoderm and endoderm are called diploblastic and the more developed animals whose structures and organs are formed from three germ layers are called triploblastic. All of a triploblastic animal's tissues and organs are derived from the three germ layers of the embryo, the ectoderm, mesoderm and endoderm.

Animal tissues can be grouped into four basic types: connective, epithelial, muscle and nervous tissue.

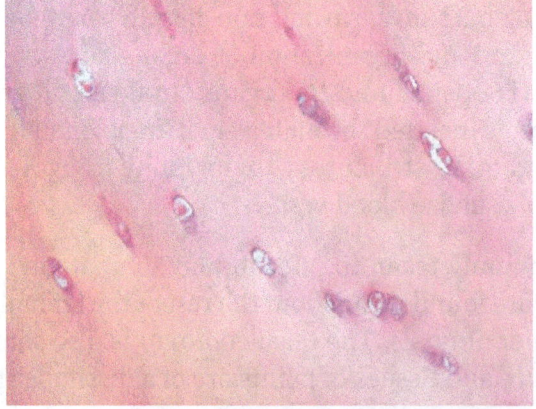

Hyaline cartilage at high magnification (H&E stain)

Connective Tissue

Connective tissues are fibrous and made up of cells scattered among inorganic material called the extracellular matrix. Connective tissue gives shape to organs and holds them in place. The main types are loose connective tissue, adipose tissue, fibrous connective tissue, cartilage and bone. The extracellular matrix contains proteins, the chief and most abundant of which is collagen. Collagen plays a major part in organizing and maintaining tissues. The matrix can be modified to form a skeleton to support or protect the body. An exoskeleton is a thickened, rigid cuticle which is stiffened by mineralization, as in crustaceans or by the cross-linking of its proteins as in insects. An endoskeleton is internal and present in all developed animals, as well as in many of those less developed.

Epithelium

Epithelial tissue is composed of closely packed cells, bound to each other by cell adhesion molecules, with little intercellular space. Epithelial cells can be squamous (flat), cuboidal or columnar and rest on a basal lamina, the upper layer of the basement membrane, the lower layer is the reticular lamina lying next to the connective tissue in the extracellular matrix secreted by the epithelial cells. There are many different types of epithelium, modified to suit a particular function. In the respiratory tract there is a type of ciliated epithelial lining; in the small intestine there are microvilli on the epithelial lining and in the large intestine there are intestinal villi. Skin consists of an outer layer of keratinized stratified squamous epithelium that covers the exterior of the vertebrate body. Keratinocytes make up to 95% of the cells in the skin. The epithelial cells on the external surface of the body typically secrete an extracellular matrix in the form of a cuticle. In simple animals this may just be a coat of glycoproteins. In more advanced animals, many glands are formed of epithelial cells.

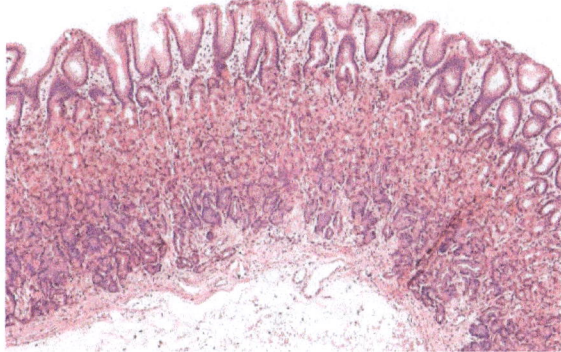

Gastric mucosa at low magnification (H&E stain)

Muscle Tissue

Muscle cells (myocytes) form the active contractile tissue of the body. Muscle tissue functions to produce force and cause motion, either locomotion or movement within internal organs. Muscle is formed of contractile filaments and is separated into three main types; smooth muscle, skeletal muscle and cardiac muscle. Smooth muscle has no striations when examined microscopically. It contracts slowly but maintains contractibility over a wide range of stretch lengths. It is found in such organs as sea anemone tentacles and the body wall of sea cucumbers. Skeletal muscle contracts rapidly but has a limited range of extension. It is found in the movement of appendages and jaws. Obliquely striated muscle is intermediate between the other two. The filaments are staggered and this is the type of muscle found in earthworms that can extend slowly or make rapid contractions. In higher animals striated muscles occur in bundles attached to bone to provide movement and are often arranged in antagonistic sets. Smooth muscle is found in the walls of the uterus, bladder, intestines, stomach, oesophagus, respiratory airways, and blood vessels. Cardiac muscle is found only in the heart, allowing it to contract and pump blood round the body.

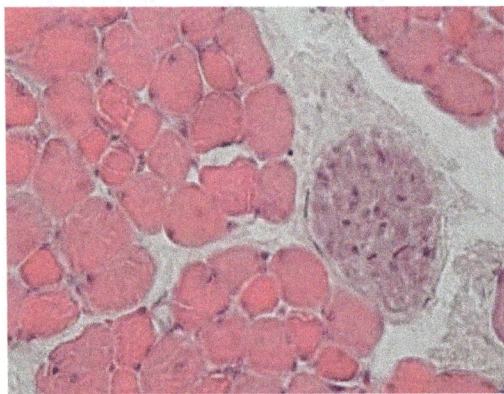

Cross section through skeletal muscle and a small nerve at high magnification (H&E stain)

Nervous Tissue

Nervous tissue is composed of many nerve cells known as neurons which transmit information. In some slow-moving radially symmetrical marine animals such as ctenophores and cnidarians (including sea anemones and jellyfish), the nerves form a nerve net, but in most animals they are organized longitudinally into bundles. In simple animals, receptor neurons in the body wall cause a local reaction to a stimulus. In more complex animals, specialized receptor cells such as chemoreceptors and photoreceptors are found in groups and send messages along neural networks to other parts of the organism. Neurons can be connected together in ganglia. In higher animals, specialized receptors are the basis of sense organs and there is a central nervous system (brain and spinal cord) and a peripheral nervous system. The latter consists of sensory nerves that transmit information from sense organs and motor nerves that influence target organs. The peripheral nervous system is divided into the somatic nervous system which conveys sensation and controls voluntary muscle, and the autonomic nervous system which involuntarily controls smooth muscle, certain glands and internal organs, including the stomach.

Vertebrate Anatomy

All vertebrates have a similar basic body plan and at some point in their lives, mostly in the embryonic stage, share the major chordate characteristics; a stiffening rod, the notochord; a dorsal hollow tube of nervous material, the neural tube; pharyngeal arches; and a tail posterior to the anus. The spinal cord is protected by the vertebral column and is above the notochord and the gastrointestinal tract is below it. Nervous tissue is derived from the ectoderm, connective tissues are derived from mesoderm, and gut is derived from the endoderm. At the posterior end is a tail which continues the spinal cord and vertebrae but not the gut. The mouth is found at the anterior end of the animal, and the anus at the base of the tail. The defining characteristic of a vertebrate is the vertebral column, formed in the development of the segmented series of vertebrae. In most vertebrates the notochord becomes the nucleus pulposus of the intervertebral discs. However, a few vertebrates, such as the sturgeon and the coelacanth retain the notochord into adulthood. Jawed vertebrates are typified by paired appendages, fins or legs, which may be secondarily lost. The limbs of vertebrates are considered to be homologous because the same underlying skeletal structure was inherited from their last common ancestor. This is one of the arguments put forward by Charles Darwin to support his theory of evolution.

Fish Anatomy

Cutaway diagram showing various organs of a fish

The body of a fish is divided into a head, trunk and tail, although the divisions between the three are not always externally visible. The skeleton, which forms the support structure inside the fish, is either made of cartilage, in cartilaginous fish, or bone in bony fish. The main skeletal element is the vertebral column, composed of articulating vertebrae which are lightweight yet strong. The ribs attach to the spine and there are no limbs or limb girdles. The main external features of the fish, the fins, are composed of either bony or soft spines called rays, which with the exception of the caudal fins, have no direct connection with the spine. They are supported by the muscles which compose the main part of the trunk. The heart has two chambers and pumps the blood through the respiratory surfaces of the gills and on round the body in a single circulatory loop. The eyes are adapted for seeing underwater and have only local vision. There is an inner ear but no external or middle ear. Low frequency vibrations are detected by the lateral line system of sense organs that run along the length of the sides of fish, and these respond to nearby movements and to changes in water pressure.

Sharks and rays are basal fish with numerous primitive anatomical features similar to those of ancient fish, including skeletons composed of cartilage. Their bodies tend to be dorso-ventrally flattened, they usually have five pairs of gill slits and a large mouth set on the underside of the head. The dermis is covered with separate dermal placoid scales. They have a cloaca into which the urinary and genital passages open, but not a swim bladder. Cartilaginous fish produce a small number of large, yolky eggs. Some species are ovoviviparous and the young develop internally but others are oviparous and the larvae develop externally in egg cases.

The bony fish lineage shows more derived anatomical traits, often with major evolutionary changes from the features of ancient fish. They have a bony skeleton, are generally laterally flattened, have five pairs of gills protected by an operculum, and a mouth at or near the tip of the snout. The dermis is covered with overlapping scales. Bony fish have a swim bladder which helps them maintain a constant depth in the water column, but not a cloaca. They mostly spawn a large number of small eggs with little yolk which they broadcast into the water column.

Amphibian Anatomy

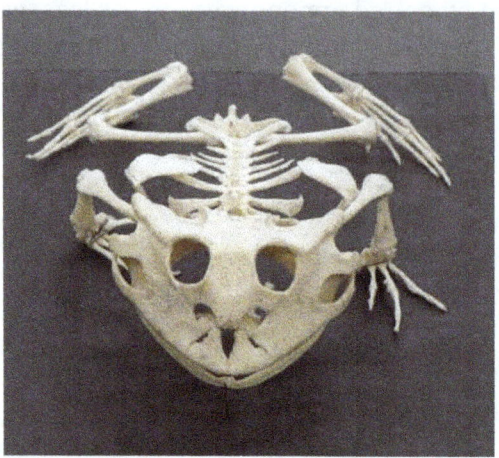

Skeleton of Surinam horned frog (*Ceratophrys cornuta*)

Amphibians are a class of animals comprising frogs, salamanders and caecilians. They are tetra-pods, but the caecilians and a few species of salamander have either no limbs or their limbs are much reduced in size. Their main bones are hollow and lightweight and are fully ossified and the vertebrae interlock with each other and have articular processes. Their ribs are usually short and may be fused to the vertebrae. Their skulls are mostly broad and short, and are often incompletely ossified. Their skin contains little keratin and lacks scales, but contains many mucous glands and in some species, poison glands. The hearts of amphibians have three chambers, two atria and one ventricle. They have a urinary bladder and nitrogenous waste products are excreted primarily as urea. Amphibians breathe by means of buccal pumping, a pump action in which air is first drawn into the buccopharyngeal region through the nostrils. These are then closed and the air is forced into the lungs by contraction of the throat. They supplement this with gas exchange through the skin which needs to be kept moist.

Plastic model of a frog

In frogs the pelvic girdle is robust and the hind legs are much longer and stronger than the fore-limbs. The feet have four or five digits and the toes are often webbed for swimming or have suction pads for climbing. Frogs have large eyes and no tail. Salamanders resemble lizards in appearance; their short legs project sideways, the belly is close to or in contact with the ground and they have a long tail. Caecilians superficially resemble earthworms and are limbless. They burrow by means of zones of muscle contractions which move along the body and they swim by undulating their body from side to side.

Reptile Anatomy

Skeleton of a diamondback rattlesnake

Reptiles are a class of animals comprising turtles, tuataras, lizards, snakes and crocodiles. They are tetrapods, but the snakes and a few species of lizard either have no limbs or their limbs are much reduced in size. Their bones are better ossified and their skeletons stronger than those of amphibians. The teeth are conical and mostly uniform in size. The surface cells of the epidermis are modified into horny scales which create a waterproof layer. Reptiles are unable to use their skin for respiration as do amphibians and have a more efficient respiratory system drawing air into their lungs by expanding their chest walls. The heart resembles that of the amphibian but there is a septum which more completely separates the oxygenated and deoxygenated bloodstreams. The reproductive system has evolved for internal fertilization, with a copulatory organ present in most species. The eggs are surrounded by amniotic membranes which prevents them from drying out and are laid on land, or develop internally in some species. The bladder is small as nitrogenous waste is excreted as uric acid.

Turtles are notable for their protective shells. They have an inflexible trunk encased in a horny carapace above and a plastron below. These are formed from bony plates embedded in the dermis which are overlain by horny ones and are partially fused with the ribs and spine. The neck is long and flexible and the head and the legs can be drawn back inside the shell. Turtles are vegetarians and the typical reptile teeth have been replaced by sharp, horny plates. In aquatic species, the front legs are modified into flippers.

Tuataras superficially resemble lizards but the lineages diverged in the Triassic period. There is one living species, *Sphenodon punctatus*. The skull has two openings (fenestrae) on either side and the jaw is rigidly attached to the skull. There is one row of teeth in the lower jaw and this fits between the two rows in the upper jaw when the animal chews. The teeth are merely projections of bony material from the jaw and eventually wear down. The brain and heart are more primitive than those of other reptiles, and the lungs have a single chamber and lack bronchi. The tuatara has a well-developed parietal eye on its forehead.

Lizards have skulls with only one fenestra on each side, the lower bar of bone below the second fenestra having been lost. This results in the jaws being less rigidly attached which allows the mouth to open wider. Lizards are mostly quadrupeds, with the trunk held off the ground by short, sideways-facing legs, but a few species have no limbs and resemble snakes. Lizards have moveable eyelids, eardrums are present and some species have a central parietal eye.

Snakes are closely related to lizards, having branched off from a common ancestral lineage during the Cretaceous period, and they share many of the same features. The skeleton consists of a skull, a hyoid bone, spine and ribs though a few species retain a vestige of the pelvis and rear limbs in the form of pelvic spurs. The bar under the second fenestra has also been lost and the jaws have extreme flexibility allowing the snake to swallow its prey whole. Snakes lack moveable eyelids, the eyes being covered by transparent "spectacle" scales. They do not have eardrums but can detect ground vibrations through the bones of their skull. Their forked tongues are used as organs of taste and smell and some species have sensory pits on their heads enabling them to locate warm-blooded prey.

Crocodilians are large, low-slung aquatic reptiles with long snouts and large numbers of teeth. The head and trunk are dorso-ventrally flattened and the tail is laterally compressed. It undulates from side to side to force the animal through the water when swimming. The tough keratinized scales provide body armour and some are fused to the skull. The nostrils, eyes and ears are elevated above the top of the flat head enabling them to remain above the surface of the water when the animal is floating. Valves seal the nostrils and ears when it is submerged. Unlike other reptiles, crocodilians have hearts with four chambers allowing complete separation of oxygenated and de-oxygenated blood.

Bird Anatomy

Part of a wing. Albrecht Dürer, c. 1500–1512

Birds are tetrapods but though their hind limbs are used for walking or hopping, their front limbs are wings covered with feathers and adapted for flight. Birds are endothermic, have a high metabolic rate, a light skeletal system and powerful muscles. The long bones are thin, hollow and very light. Air sac extensions from the lungs occupy the centre of some bones. The sternum is wide and usually has a keel and the caudal vertebrae are fused. There are no teeth and the narrow jaws are adapted into a horn-covered beak. The eyes are relatively large, particularly in nocturnal species such as owls. They face forwards in predators and sideways in ducks.

The feathers are outgrowths of the epidermis and are found in localized bands from where they fan out over the skin. Large flight feathers are found on the wings and tail, contour feathers cover the bird's surface and fine down occurs on young birds and under the contour feathers of water birds. The only cutaneous gland is the single uropygial gland near the base of the tail. This produces an oily secretion that waterproofs the feathers when the bird preens. There are scales on the legs, feet and claws on the tips of the toes.

Mammal Anatomy

Mammals are a diverse class of animals, mostly terrestrial but some are aquatic and others have evolved flapping or gliding flight. They mostly have four limbs but some aquatic mammals have no limbs or limbs modified into fins and the forelimbs of bats are modified into wings. The legs of most mammals are situated below the trunk, which is held well clear of the ground. The bones of mammals are well ossified and their teeth, which are usually differentiated, are coated in a layer of prismatic enamel. The teeth are shed once (milk teeth) during the animal's lifetime or not at all, as is the case in cetaceans. Mammals have three bones in the middle ear and a cochlea in the inner ear. They are clothed in hair and their skin contains glands which secrete sweat. Some of these glands are specialized as mammary glands, producing milk to feed the young. Mammals breathe with lungs and have a muscular diaphragm separating the thorax from the abdomen which helps them draw air into the lungs. The mammalian heart has four chambers and oxygenated and deoxygenated blood are kept entirely separate. Nitrogenous waste is excreted primarily as urea.

Mammals are amniotes, and most are viviparous, giving birth to live young. The exception to this are the egg-laying monotremes, the platypus and the echidnas of Australia. Most other mammals have a placenta through which the developing foetus obtains nourishment, but in marsupials, the foetal stage is very short and the immature young is born and finds its way to its mother's pouch where it latches on to a nipple and completes its development.

Invertebrate Anatomy

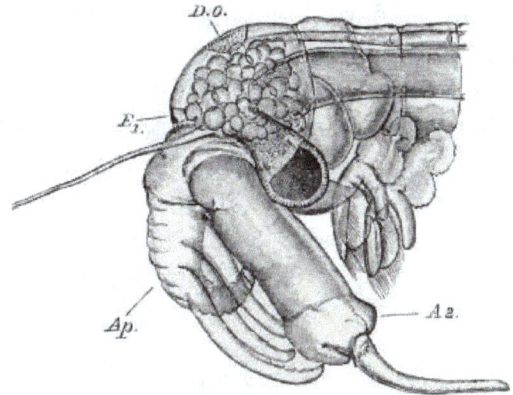

Head of a male *Daphnia*, a planktonic crustacean

Invertebrates constitute a vast array of living organisms ranging from the simplest unicellular eukaryotes such as *Paramecium* to such complex multicellular animals as the octopus, lobster and dragonfly. They constitute about 95% of the animal species. By definition, none of these creatures has a backbone. The cells of single-cell protozoans have the same basic structure as those of multicellular animals but some parts are specialized into the equivalent of tissues and organs. Locomotion is often provided by cilia or flagella or may proceed via the advance of pseudopodia, food may be gathered by phagocytosis, energy needs may be supplied by photosynthesis and the cell may be supported by an endoskeleton or an exoskeleton. Some protozoans can form multicellular colonies.

Metazoans are multicellular organism, different groups of cells of which have separate functions. The most basic types of metazoan tissues are epithelium and connective tissue, both of which are present in nearly all invertebrates. The outer surface of the epidermis is normally formed of epithelial cells and secretes an extracellular matrix which provides support to the organism. An endoskeleton derived from the mesoderm is present in echinoderms, sponges and some cephalopods. Exoskeletons are derived from the epidermis and is composed of chitin in arthropods (insects, spiders, ticks, shrimps, crabs, lobsters). Calcium carbonate constitutes the shells of molluscs, brachiopods and some tube-building polychaete worms and silica forms the exoskeleton of the microscopic diatoms and radiolaria. Other invertebrates may have no rigid structures but the epidermis may secrete a variety of surface coatings such as the pinacoderm of sponges, the gelatinous cuticle of cnidarians (polyps, sea anemones, jellyfish) and the collagenous cuticle of annelids. The outer epithelial layer may include cells of several types including sensory cells, gland cells and stinging cells. There may also be protrusions such as microvilli, cilia, bristles, spines and tubercles.

Marcello Malpighi, the father of microscopical anatomy, discovered that plants had tubules similar to those he saw in insects like the silk worm. He observed that when a ring-like portion of bark was removed on a trunk a swelling occurred in the tissues above the ring, and he unmistakably interpreted this as growth stimulated by food coming down from the leaves, and being captured above the ring.

Arthropod Anatomy

Arthropods comprise the largest phylum in the animal kingdom with over a million known invertebrate species.

Insects possess segmented bodies supported by a hard-jointed outer covering, the exoskeleton, made mostly of chitin. The segments of the body are organized into three distinct parts, a head, a thorax and an abdomen. The head typically bears a pair of sensory antennae, a pair of compound eyes, one to three simple eyes (ocelli) and three sets of modified appendages that form the mouthparts. The thorax has three pairs of segmented legs, one pair each for the three segments that compose the thorax and one or two pairs of wings. The abdomen is composed of eleven segments, some of which may be fused and houses the digestive, respiratory, excretory and reproductive systems. There is considerable variation between species and many adaptations to the body parts, especially wings, legs, antennae and mouthparts.

Spiders a class of arachnids have four pairs of legs; a body of two segments—a cephalothorax and an abdomen. Spiders have no wings and no antennae. They have mouthparts called chelicerae which are often connected to venom glands as most spiders are venomous. They have a second pair of appendages called pedipalps attached to the cephalothorax. These have similar segmentation to the legs and function as taste and smell organs. At the end of each male pedipalp is a spoon-shaped cymbium that acts to support the copulatory organ.

Other Branches of Anatomy

- Superficial or surface anatomy is important as the study of anatomical landmarks that can be readily seen from the exterior contours of the body. It enables physicians or veterinary surgeons to gauge the position and anatomy of the associated deeper structures. Superficial is a directional term that indicates that structures are located relatively close to the surface of the body.

- Comparative anatomy relates to the comparison of anatomical structures (both gross and microscopic) in different animals.

- Artistic anatomy relates to anatomic studies for artistic reasons.

Fish Anatomy

Fish anatomy is the study of the form or morphology of fishes. It can be contrasted with fish physiology, which is the study of how the component parts of fish function together in the living fish. In practice, fish anatomy and fish physiology complement each other, the former dealing with the structure of a fish, its organs or component parts and how they are put together, such as might be observed on the dissecting table or under the microscope, and the latter dealing with how those components function together in living fish.

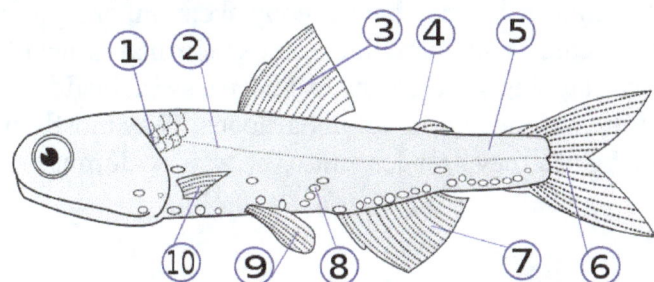

External anatomy of a bony fish (Hector's lanternfish)
1. operculum (gill cover), 2. lateral line, 3. dorsal fin, 4. adipose fin, 5. caudal peduncle, 6. caudal fin
7. anal fin, 8. photophores, 9. pelvic fins (paired), 10. pectoral fins (paired)

The anatomy of fish is often shaped by the physical characteristics of water, the medium in which fish live. Water is much denser than air, holds a relatively small amount of dissolved oxygen, and absorbs more light than air does. The body of a fish is divided into a head, trunk and tail, although the divisions between the three are not always externally visible. The skeleton, which forms the support structure inside the fish, is either made of cartilage, in cartilaginous fish, or bone in bony fish. The main skeletal element is the vertebral column, composed of articulating vertebrae which are lightweight yet strong. The ribs attach to the spine and there are no limbs or limb girdles. The main external features of the fish, the fins, are composed of either bony or soft spines called rays which, with the exception of the caudal fins, have no direct connection with the spine. They are supported by the muscles which compose the main part of the trunk. The heart has two chambers and pumps the blood through the respiratory surfaces of the gills and on round the body in a single circulatory loop. The eyes are adapted for seeing underwater and have only local vision. There is an inner ear but no external or middle ear. Low frequency vibrations are detected by the lateral line system of sense organs that run along the length of the sides of fish, and these respond to nearby movements and to changes in water pressure.

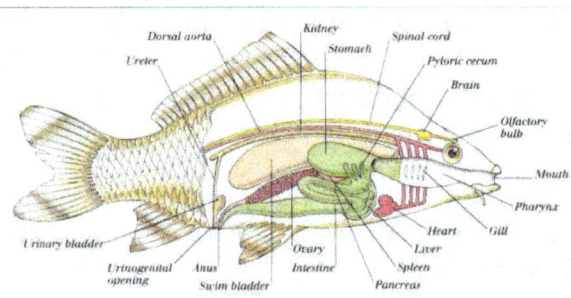

Internal anatomy of a bony fish

Sharks and rays are basal fish with numerous primitive anatomical features similar to those of ancient fish, including skeletons composed of cartilage. Their bodies tend to be dorso-ventrally flattened, they usually have five pairs of gill slits and a large mouth set on the underside of the head. The dermis is covered with separate dermal placoid scales. They have a cloaca into which the urinary and genital passages open, but not a swim bladder. Cartilaginous fish produce a small number of large, yolky eggs. Some species are ovoviviparous and the young develop internally but others are oviparous and the larvae develop externally in egg cases.

The bony fish lineage shows more derived anatomical traits, often with major evolutionary changes from the features of ancient fish. They have a bony skeleton, are generally laterally flattened, have five pairs of gills protected by an operculum, and a mouth at or near the tip of the snout. The dermis is covered with overlapping scales. Bony fish have a swim bladder which helps them maintain a constant depth in the water column, but not a cloaca. They mostly spawn a large number of small eggs with little yolk which they broadcast into the water column.

Body

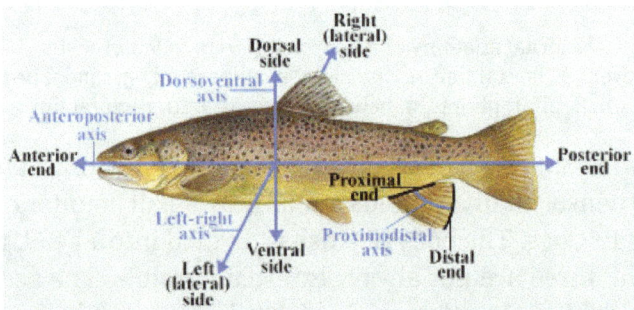

Anatomical directions and axes

In many respects fish anatomy is different from humans and mammals, yet it shares the same basic vertebrate body plan from which all vertebrates have evolved: a notochord, rudimentary vertebrae, and a well-defined head and tail.

Fish have a variety of different body plans. At the broadest level their body is divided into head, trunk, and tail, although the divisions are not always externally visible. The body is often fusiform, a streamlined body plan often found in fast-moving fish. They may also be filiform (eel-shaped) or vermiform (worm-shaped). Also, fish are often either compressed (laterally thin) or depressed (dorso-ventrally flat).

Skeleton

Skeleton of a bony fish

There are two different skeletal types: the exoskeleton, which is the stable outer shell of an organism, and the endoskeleton, which forms the support structure inside the body. The skeleton of the fish is either made of cartilage (cartilaginous fishes) or bones (bony fishes). The main features of the fish, the fins, are bony fin rays and, except for the caudal fin, have no direct connection with the spine. They are supported only by the muscles. The ribs attach to the spine.

Bones are rigid organs that form part of the endoskeleton of vertebrates. They function to move, support, and protect the various organs of the body, produce red and white blood cells and store minerals. Bone tissue is a type of dense connective tissue. Because bones come in a variety of shapes and have a complex internal and external structure they are lightweight, yet strong and hard, in addition to fulfilling their many other functions.

Fish bones have been used to bioremediate lead from contaminated soil.

Vertebrae

Skeletal structure of a bass showing the vertebral column running from the head to the tail

Skeletal structure of an Atlantic cod

Fish are vertebrates. All vertebrates are built along the basic chordate body plan: a stiff rod running through the length of the animal (vertebral column or notochord), with a hollow tube of nervous tissue (the spinal cord) above it and the gastrointestinal tract below. In all vertebrates, the mouth is found at, or right below, the anterior end of the animal, while the anus opens to the exterior before the end of the body. The remaining part of the body continuing aft of the anus forms a tail with vertebrae and spinal cord, but no gut.

The defining characteristic of a vertebrate is the vertebral column, in which the notochord (a stiff rod of uniform composition) found in all chordates has been replaced by a segmented series of stiffer elements (vertebrae) separated by mobile joints (intervertebral discs, derived embryonically and evolutionarily from the notochord). However, a few fish have secondarily lost this anatomy, retaining the notochord into adulthood, such as the sturgeon.

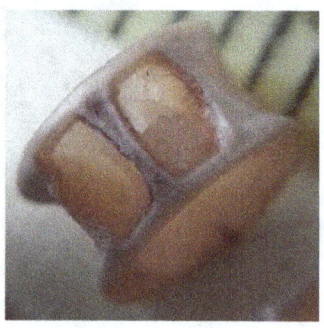

The X-ray tetra (*Pristella maxillaris*) has a visible back-bone. The spinal cord is housed within its backbone

A vertebra (diameter 5 mm) of a small ray-finned fish

The vertebral column consists of a centrum (the central body or spine of the vertebra), vertebral arches which protrude from the top and bottom of the centrum, and various processes which project from the centrum or arches. An arch extending from the top of the centrum is called a neural arch, while the hemal arch or chevron is found underneath the centrum in the caudal (tail) vertebrae of fish. The centrum of a fish is usually concave at each end (amphicoelous), which limits the motion of the fish. This can be contrasted with the centrum of a mammal, which is flat at each end (acoelous), shaped in a manner that can support and distribute compressive forces.

The vertebrae of lobe-finned fishes consist of three discrete bony elements. The vertebral arch surrounds the spinal cord, and is of broadly similar form to that found in most other vertebrates. Just beneath the arch lies a small plate-like *pleurocentrum*, which protects the upper surface of the notochord, and below that, a larger arch-shaped *intercentrum* to protect the lower border. Both of these structures are embedded within a single cylindrical mass of cartilage. A similar arrangement was found in primitive tetrapods, but, in the evolutionary line that led to reptiles (and hence, also to mammals and birds), the intercentrum became partially or wholly replaced by an enlarged pleurocentrum, which in turn became the bony vertebral body.

In most ray-finned fishes, including all teleosts, these two structures are fused with, and embedded within, a solid piece of bone superficially resembling the vertebral body of mammals. In living amphibians, there is simply a cylindrical piece of bone below the vertebral arch, with no trace of the separate elements present in the early tetrapods.

In cartilagenous fish, such as sharks, the vertebrae consist of two cartilagenous tubes. The upper tube is formed from the vertebral arches, but also includes additional cartilagenous structures filling in the gaps between the vertebrae, and so enclosing the spinal cord in an essentially continuous sheath. The lower tube surrounds the notochord, and has a complex structure, often including multiple layers of calcification.

Lampreys have vertebral arches, but nothing resembling the vertebral bodies found in all higher vertebrates. Even the arches are discontinuous, consisting of separate pieces of arch-shaped cartilage around the spinal cord in most parts of the body, changing to long strips of cartilage above and below in the tail region. Hagfishes lack a true vertebral column, and are therefore not properly considered vertebrates, but a few tiny neural arches are present in the tail. Hagfishes do, however, possess a cranium. For this reason, the vertebrate subphylum is sometimes referred to as "Craniata" when discussing morphology. Molecular analysis since 1992 has suggested that the hagfishes are most closely related to lampreys, and so also are vertebrates in a monophyletic sense. Others consider them a sister group of vertebrates in the common taxon of Craniata.

Head

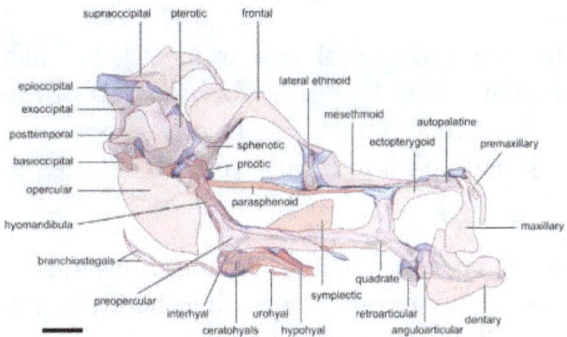

Skull bones as they appear in a seahorse

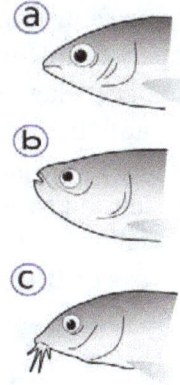

Positions of fish mouths:
(a) - terminal, (b) - superior, (c) - subterminal, inferior

The head or skull includes the skull roof (a set of bones covering the brain, eyes and nostrils), the snout (from the eye to the forward most point of the upper jaw), the operculum or gill cover (absent in sharks and jawless fish), and the cheek, which extends from the eye to preopercle. The operculum and preopercle may or may not have spines. In sharks and some primitive bony fish a spiracle, small extra gill opening, is found behind each eye.

The skull in fishes is formed from a series of only loosely connected bones. Jawless fish and sharks only possess a cartilaginous endocranium, with the upper and lower jaws of cartilaginous fish being separate elements not attached to the skull.Bony fishes have additional dermal bone, forming a more or less coherent skull roof in lungfish and holost fish. The lower jaw defines a chin.

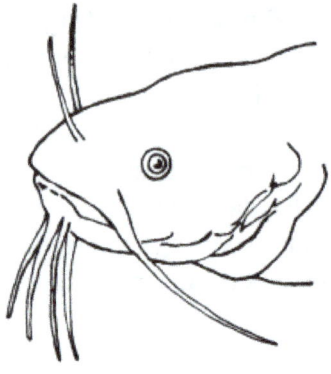

Barbels

In lampreys, the mouth is formed into an oral disk. In most jawed fish, however, there are three general configurations. The mouth may be on the forward end of the head (*terminal*), may be upturned (*superior*), or may be turned downwards or on the bottom of the fish (*subterminal* or *inferior*). The mouth may be modified into a suckermouth adapted for clinging onto objects in fast-moving water.

The simpler structure is found in jawless fish, in which the cranium is represented by a trough-like basket of cartilaginous elements only partially enclosing the brain, and associated with the capsules for the inner ears and the single nostril. Distinctively, these fish have no jaws.

Cartilaginous fish, such as sharks, also have simple, and presumably primitive, skull structures. The cranium is a single structure forming a case around the brain, enclosing the lower surface and the sides, but always at least partially open at the top as a large fontanelle. The most anterior part of the cranium includes a forward plate of cartilage, the rostrum, and capsules to enclose the olfactory organs. Behind these are the orbits, and then an additional pair of capsules enclosing the structure of the inner ear. Finally, the skull tapers towards the rear, where the foramen magnum lies immediately above a single condyle, articulating with the first vertebra. There are, in addition, at various points throughout the cranium, smaller foramina for the cranial nerves. The jaws consist of separate hoops of cartilage, almost always distinct from the cranium proper.

Skull of a northern pike

In the ray-finned fishes, there has also been considerable modification from the primitive pattern. The roof of the skull is generally well formed, and although the exact relationship of its bones to those of tetrapods is unclear, they are usually given similar names for convenience. Other elements of the skull, however, may be reduced; there is little cheek region behind the enlarged orbits, and little, if any bone in between them. The upper jaw is often formed largely from the premaxilla, with the maxilla itself located further back, and an additional bone, the symplectic, linking the jaw to the rest of the cranium.

Although the skulls of fossil lobe-finned fish resemble those of the early tetrapods, the same cannot be said of those of the living lungfishes. The skull roof is not fully formed, and consists of multiple, somewhat irregularly shaped bones with no direct relationship to those of tetrapods. The upper jaw is formed from the pterygoids and vomers alone, all of which bear teeth. Much of the skull is formed from cartilage, and its overall structure is reduced.

Skull of *Tiktaalik*, a genus of extinct sarcopterygian (lobe-finned "fish") from the late Devonian period

The head may have several fleshy structures known as barbels, which may be very long and resemble whiskers. Many fish species also have a variety of protrusions or spines on the head. The nostrils or nares of almost all fishes do not connect to the oral cavity, but are pits of varying shape and depth.

External Organs

Jaw

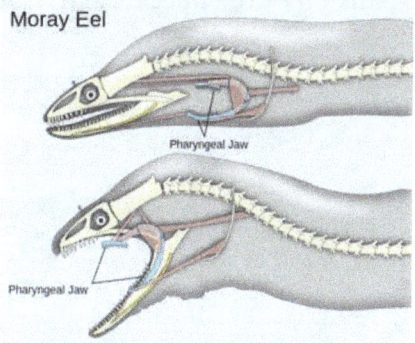

Moray eels have two sets of jaws: the oral jaws that capture prey and the pharyngeal jaws that advance into the mouth and move prey from the oral jaws to the esophagus for swallowing.

The vertebrate jaw probably originally evolved in the Silurian period and appeared in the Placoderm fish which further diversified in the Devonian. Jaws are thought to derive from the pharyngeal arches that support the gills in fish. The two most anterior of these arches are thought to have become the jaw itself and the hyoid arch, which braces the jaw against the braincase and increases mechanical efficiency. While there is no fossil evidence directly to support this theory, it makes sense in light of the numbers of pharyngeal arches that are visible in extant jawed (the Gnathostomes), which have seven arches, and primitive jawless vertebrates (the Agnatha), which have nine.

Jaws of great white shark

It is thought that the original selective advantage garnered by the jaw was not related to feeding, but to increased respiration efficiency. The jaws were used in the buccal pump (observable in modern fish and amphibians) that pumps water across the gills of fish or air into the lungs in the case of amphibians. Over evolutionary time the more familiar use of jaws (to humans), in feeding, was selected for and became a very important function in vertebrates.

Linkage systems are widely distributed in animals. The most thorough overview of the different types of linkages in animals has been provided by M. Muller, who also designed a new classification system, which is especially well suited for biological systems. Linkage mechanisms are especially frequent and manifold in the head of bony fishes, such as wrasses, which have evolved many specialized feeding mechanisms. Especially advanced are the linkage mechanisms of jaw protrusion. For suction feeding a system of linked four-bar linkages is responsible for the coordinated opening of the mouth and 3-D expansion of the buccal cavity. Other linkages are responsible for protrusion of the premaxilla.

Eyes

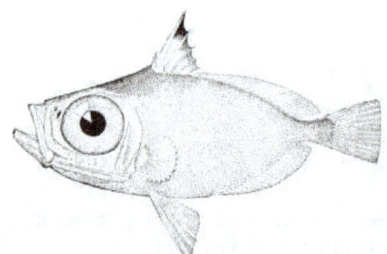

Zenion hololepis is a small deep water fish
with large eyes

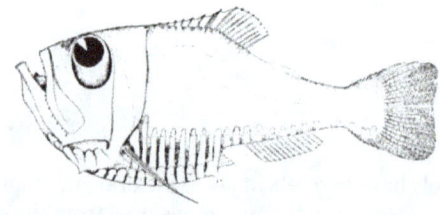

The deep sea half-naked hatchetfish has eyes which look
overhead where it can see the silhouettes of prey

Fish eyes are similar to terrestrial vertebrates like birds and mammals, but have a more spherical lens. Their retinas generally have both rod cells and cone cells (for scotopic and photopic vision), and most species have colour vision. Some fish can see ultraviolet and some can see polarized light. Amongst jawless fish, the lamprey has well-developed eyes, while the hagfish has only primitive eyespots. The ancestors of modern hagfish, thought to be the protovertebrate were evidently pushed to very deep, dark waters, where they were less vulnerable to sighted predators, and where it is advantageous to have a convex eye-spot, which gathers more light than a flat or concave one. Unlike humans, fish normally adjust focus by moving the lens closer to or further from the retina.

Gills

Gill of a rainbow trout

The gills, located under the operculum, are a respiratory organ for the extraction of oxygen from water and for the excretion of carbon dioxide. They are not usually visible, but can be seen in some species, such as the frilled shark. The labyrinth organ of Anabantoidei and Clariidae is used to allow the fish to extract oxygen from the air. Gill rakers are bony or cartilaginous, finger-like projections off the gill arch which function in filter-feeders to retain filtered prey.

Skin

The epidermis of fish consists entirely of live cells, with only minimal quantities of keratin in the cells of the superficial layer. It is generally permeable. The dermis of bony fish typically contains relatively little of the connective tissue found in tetrapods. Instead, in most species, it is largely replaced by solid, protective bony scales. Apart from some particularly large dermal bones that form parts of the skull, these scales are lost in tetrapods, although many reptiles do have scales of a different kind, as do pangolins. Cartilaginous fish have numerous tooth-like denticles embedded in their skin, in place of true scales.

Sweat glands and sebaceous glands are both unique to mammals, but other types of skin glands are found in fish. Fish typically have numerous individual mucus-secreting skin cells that aid in insulation and protection, but may also have poison glands, photophores, or cells that produce a more watery, serous fluid. Melanin colours the skin of many species, but in fish the epidermis is often relatively colourless. Instead, the colour of the skin is largely due to chromatophores in the dermis, which, in addition to melanin, may contain guanine or carotenoid pigments. Many species, such as flounders, change the colour of their skin by adjusting the relative size of their chromatophores.

Scales

Cycloid scales covering rohu

The outer body of many fish is covered with scales, which are part of the fish's integumentary system. The scales originate from the mesoderm (skin), and may be similar in structure to teeth. Some species are covered instead by scutes. Others have no outer covering on the skin. Most fish are covered in a protective layer of slime (mucus).

There are four principal types of fish scales.

1. Placoid scales, also called dermal denticles, are similar to teeth in that they are made of dentin covered by enamel. They are typical of sharks and rays.

2. Ganoid scales are flat, basal-looking scales that cover a fish body with little overlapping. They are typical of gar and bichirs.

3. Cycloid scales are small oval-shaped scales with growth rings. Bowfin and remora have cycloid scales.

4. Ctenoid scales are similar to the cycloid scales, with growth rings. They are distinguished by spines that cover one edge. Halibut have this type of scale.

Another, less common, type of scale is the scute, which is:

- an external shield-like bony plate, or

- a modified, thickened scale that often is keeled or spiny, or

- a projecting, modified (rough and strongly ridged) scale, usually associated with the lateral line, or on the caudal peduncle forming caudal keels, or along the ventral profile. Some fish, such as pineconefish, are completely or partially covered in scutes.

Lateral Line

The lateral line is clearly visible as a line of receptors running along the side of this Atlantic cod

The lateral line is a sense organ used to detect movement and vibration in the surrounding water. For example, fish can use their lateral line system to follow the vortices produced by fleeing prey. In most species, it consists of a line of receptors running along each side of the fish.

Photophores

Photophores are light-emitting organs which appears as luminous spots on some fishes. The light can be produced from compounds during the digestion of prey, from specialized mitochondrial cells in the organism called photocytes, or associated with symbiotic bacteria, and are used for attracting food or confusing predators.

Fins

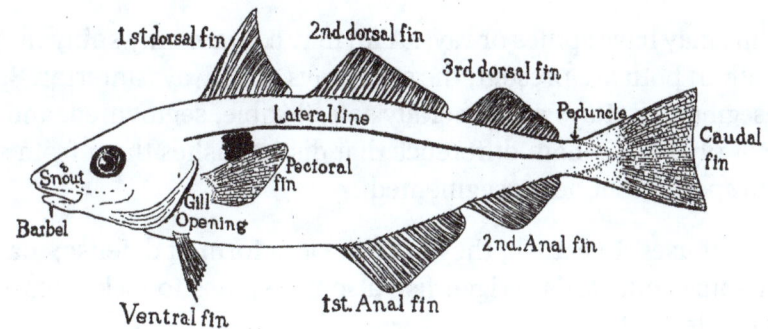

The haddock, a type of cod, is ray-finned. It has three dorsal and two anal fins

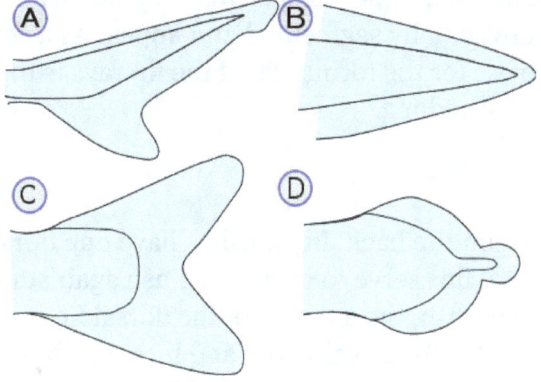

Types of caudal (tail) fin:
(A) - Heterocercal, (B) - Protocercal,
(C) - Homocercal, (D) - Diphycercal

Fins are the most distinctive features of fish. They are either composed of bony spines or rays protruding from the body with skin covering them and joining them together, either in a webbed fashion as seen in most bony fish or similar to a flipper as seen in sharks. Apart from the tail or caudal fin, fins have no direct connection with the spine and are supported by muscles only. Their principal function is to help the fish swim. Fins can also be used for gliding or crawling, as seen in the flying fish and frogfish. Fins located in different places on the fish serve different purposes, such as moving forward, turning, and keeping an upright position. For every fin, there are a number of fish species in which this particular fin has been lost during evolution.

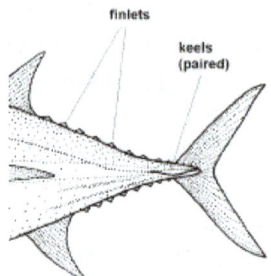

Sharks possess a heterocercal caudal fin. The dorsal portion is usually larger than the ventral portion

The high performance bigeye tuna is equipped with a
homocercal caudal fin and finlets and keels

Spines and Rays

In bony fish, most fins may have spines or rays. A fin may contain only spiny rays, only soft rays, or a combination of both. If both are present, the spiny rays are always anterior. Spines are generally stiff, sharp and unsegmented. Rays are generally soft, flexible, segmented, and may be branched. This segmentation of rays is the main difference that distinguishes them from spines; spines may be flexible in certain species, but never segmented.

Spines have a variety of uses. In catfish, they are used as a form of defense; many catfish have the ability to lock their spines outwards. Triggerfish also use spines to lock themselves in crevices to prevent them being pulled out.

Lepidotrichia are bony, bilaterally-paired, segmented fin rays found in bony fishes. They develop around actinotrichia as part of the dermal exoskeleton. Lepidotrichia may have some cartilage or bone in them as well. They are actually segmented and appear as a series of disks stacked one on top of another. The genetic basis for the formation of the fin rays is thought to be genes coding for the proteins actinodin 1 and actinodin 2.

Types of Fin

- Dorsal fins are located on the back. Most fishes have one dorsal fin, but some fishes have two or three . The dorsal fins serve to protect the fish against rolling, and assists in sudden turns and stops. In anglerfish, the anterior of the dorsal fin is modified into an *illicium* and *esca*, a biological equivalent to a fishing rod and lure. The bones that support the dorsal fin are called *Pterygiophore*. There are two to three of them: "proximal", "middle", and "distal". In spinous fins the distal is often fused to the middle, or not present at all.

- The caudal fin is the tail fin, located at the end of the caudal peduncle and is used for propulsion. The caudal peduncle is the narrow part of the fish's body to which the caudal or tail fin is attached. The hypural joint is the joint between the caudal fin and the last of the vertebrae. The hypural is often fan-shaped. The tail is called:

  o *Heterocercal* if the vertebrae extend into the upper lobe of the tail, making it longer (as in sharks)

  o *Reversed heterocercal* if the vertebrae extend into the lower lobe of the tail, making it longer (as in the Anaspida) o   *Protocercal* if the vertebrae extend to the tip of the tail and the tail is symmetrical but not expanded (as in amphioxus)

o   *Diphycercal* if the vertebrae extend to the tip of the tail and the tail is symmetrical and expanded (as in the bichir, lungfish, lamprey and coelacanth. Most Palaeozoic fishes had a diphycercal heterocercal tail.)

o   Most fish have a *homocercal* tail, where the fin appears superficially symmetric but the vertebrae extend for a very short distance into the upper lobe of the fin. This can be expressed in a variety of shapes. The tail fin can be:

   ☐   rounded at the end

   ☐   truncated: or end in a more-or-less vertical edge, such as in salmon

   ☐   forked: or end in two prongs

   ☐   emarginate: or with a slight inward curve.

   ☐   continuous: with dorsal, caudal and anal fins attached, such as in eels

- The anal fin is located on the ventral surface behind the anus. This fin is used to stabilize the fish while swimming.

- The paired pectoral fins are located on each side, usually just behind the operculum, and are homologous to the forelimbs of tetrapods. A peculiar function of pectoral fins, highly developed in some fish, is the creation of the dynamic lifting force that assists some fish, such as sharks, in maintaining depth and also enables the "flight" for flying fish. In many fish, the pectoral fins aid in walking, especially in the lobe-like fins of some anglerfish and in the mudskipper. Certain rays of the pectoral fins may be adapted into finger-like projections, such as in sea robins and flying gurnards. The "horns" of manta rays and their relatives are called *cephalic fins*; this is actually a modification of the anterior portion of the pectoral fin.

- The paired pelvic or ventral fins are located ventrally below the pectoral fins. They are homologous to the hindlimbs of tetrapods. The pelvic fin assists the fish in going up or down through the water, turning sharply, and stopping quickly. In gobies, the pelvic fins are often fused into a single sucker disk. This can be used to attach to objects.

- The adipose fin is a soft, fleshy fin found on the back behind the dorsal fin and just forward of the caudal fin. It is absent in many fish families, but is found in Salmonidae, characins and catfishes. Its function has remained a mystery, and is frequently clipped off to mark hatchery-raised fish, though data from 2005 showed that trout with their adipose fin removed have an 8% higher tailbeat frequency. Additional research published in 2011 has suggested that the fin may be vital for the detection of and response to stimuli such as touch, sound and changes in pressure. Canadian researchers identified a neural network in the fin, indicating that it likely has a sensory function, but are still not sure exactly what the consequences of removing it are.

- Some types of fast-swimming fish have a horizontal caudal keel just forward of the tail fin. Much like the keel of a ship, this is a lateral ridge on the caudal peduncle, usually composed of scutes, that provides stability and support to the caudal fin. There may be a single paired keel, one on each side, or two pairs above and below.

- Finlets are small fins, generally between the dorsal and the caudal fins also between the anal fin and the caudal fin (in bichirs, there are only finlets on the dorsal surface and no dorsal fin). In some fish such as tuna or sauries, they are rayless, non-retractable, and found between the last dorsal and/or anal fin and the caudal fin.

Internal Organs

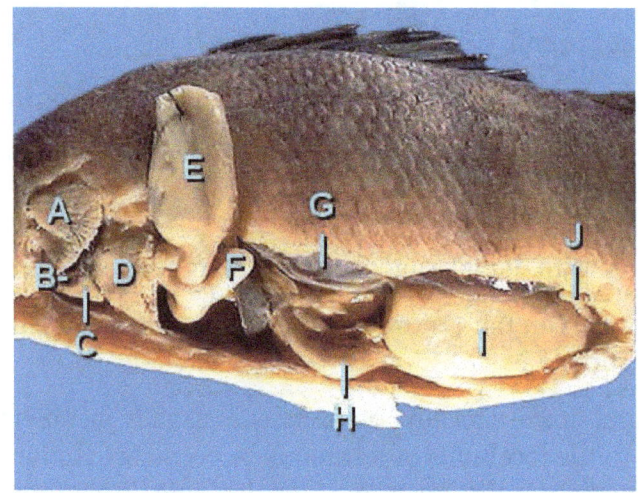

Internal organs of a male yellow perch
A = gill, B = heart atrium, C = heart ventricle, D=liver (cut), E = stomach
F = pyloric caeca, G = swim bladder, H = intestine, I = testis, J = urinary bladder

Intestines

As with other vertebrates, the intestines of fish consist of two segments, the small intestine and the large intestine. In most higher vertebrates, the small intestine is further divided into the duodenum and other parts. In fish, the divisions of the small intestine are not as clear, and the terms *anterior intestine* or *proximal intestine* may be used instead of duodenum. In bony fish, the intestine is relatively short, typically around one and a half times the length of the fish's body. It commonly has a number of *pyloric caeca*, small pouch-like structures along its length that help to increase the overall surface area of the organ for digesting food. There is no ileocaecal valve in teleosts, with the boundary between the small intestine and the rectum being marked only by the end of the digestive epithelium. There is no small intestine as such in non-teleost fish, such as sharks, sturgeons, and lungfish. Instead, the digestive part of the gut forms a *spiral intestine*, connecting the stomach to the rectum. In this type of gut, the intestine itself is relatively straight, but has a long fold running along the inner surface in a spiral fashion, sometimes for dozens of turns. This fold creates a valve-like structure that greatly increases both the surface area and the effective length of the intestine. The lining of the spiral intestine is similar to that of the small intestine in teleosts and non-mammalian tetrapods. In lampreys, the spiral valve is extremely small, possibly because their diet requires little digestion. Hagfish have no spiral valve at all, with digestion occurring for almost the entire length of the intestine, which is not subdivided into different regions.

Pyloric Caeca

The black swallower is a species of deep sea fish with an extensible stomach
which allows it to swallow fish larger than itself

The pyloric caecum is a pouch, usually peritoneal, at the beginning of the large intestine. It receives faecal material from the ileum, and connects to the ascending colon of the large intestine. It is present in most amniotes, and also in lungfish. Many fish in addition have a number of small outpocketings, also called pyloric caeca, along their intestine; despite the name they are not homologous with the caecum of amniotes, and their purpose is to increase the overall area of the digestive epithelium.

Internal organs of a female Atlantic cod
1. Liver, 2. Gas bladder, 3. ovary,
4. Pyloric caeca, 5. Stomach, 6. Intestine

Stomach

As with other vertebrates, the relative positions of the esophageal and duodenal openings to the stomach remain relatively constant. As a result, the stomach always curves somewhat to the left before curving back to meet the pyloric sphincter. However, lampreys, hagfishes, chimaeras, lungfishes, and some teleost fish have no stomach at all, with the esophagus opening directly into the intestine. These fish consume diets that either require little storage of food, or no pre-digestion with gastric juices, or both.

Kidneys

The kidneys of fish are typically narrow, elongated organs, occupying a significant portion of the trunk. They are similar to the mesonephros of higher vertebrates (reptiles, birds and mammals). The kidneys contain clusters of nephrons, serviced by collecting ducts which usually drain into a mesonephric duct. However, the situation is not always so simple. In cartilaginous fish there is also a shorter duct which drains the posterior (metanephric) parts of the kidney, and joins with the mesonephric duct at the bladder or cloaca. Indeed, in many cartilaginous fish, the anterior portion of the kidney may degenerate or cease to function altogether in the adult. Hagfish and lamprey kidneys are unusually simple. They consist of a row of nephrons, each emptying directly into the mesonephric duct.

Spleen

The spleen is found in nearly all vertebrates. It is a non-vital organ, similar in structure to a large lymph node. It acts primarily as a blood filter, and plays important roles in regard to red blood cells and the immune system. In cartilaginous and bony fish it consists primarily of red pulp and is normally a somewhat elongated organ as it actually lies inside the serosal lining of the intestine. The only vertebrates lacking a spleen are the lampreys and hagfishes. Even in these animals, there is a diffuse layer of haematopoeitic tissue within the gut wall, which has a similar structure to red pulp, and is presumed to be homologous with the spleen of higher vertebrates.

Liver

The liver is a large vital organ present in all fish. It has a wide range of functions, including detoxification, protein synthesis, and production of biochemicals necessary for digestion.

Heart

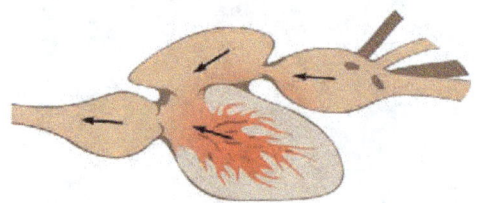

Blood flow through the heart: sinus venosus, atrium, ventricle, and outflow tract

Fish have what is often described as a two-chambered heart, consisting of one atrium to receive blood and one ventricle to pump it, in contrast to three chambers (two atria, one ventricle) of amphibian and most reptile hearts and four chambers (two atria, two ventricles) of mammal and bird hearts. However, the fish heart has entry and exit compartments that may be called chambers, so it is also sometimes described as three-chambered or four-chambered, depending on what is counted as a chamber. The atrium and ventricle are sometimes considered "true chambers", while the others are considered "accessory chambers".

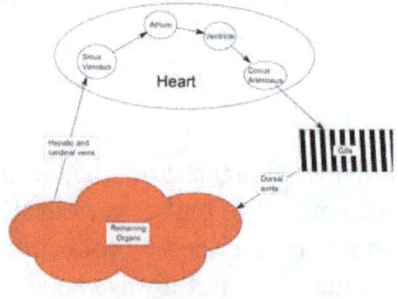

Cardiovascular cycle in a fish

The four compartments are arranged sequentially:

- Sinus venosus, a thin-walled sac or reservoir with some cardiac muscle that collects deoxygenated blood through the incoming hepatic and cardinal veins.

- Atrium, a thicker-walled, muscular chamber that sends blood to the ventricle.

- Ventricle, a thick-walled, muscular chamber that pumps the blood to the fourth part, the outflow tract. The shape of the ventricle varies considerably, usually tubular in fish with elongated bodies, pyramidal with a triangular base in others, or sometimes sac-like in some marine fish.

- The outflow tract (OFT) to the ventral aorta, consisting of the tubular conus arteriosus, bulbus arteriosus, or both. The conus arteriosus, typically found in more primitive species of fish, contracts to assist blood flow to the aorta, while the bulbus anteriosus does not.

Ostial valves, consisting of flap-like connective tissues, prevent blood from flowing backward through the compartments. The ostial valve between the sinus venosus and atrium is called the sino-atrial valve, which closes during ventricular contraction. Between the atrium and ventricle is an ostial valve called the atrio-ventricular valve, and between the bulbus arteriosus and ventricle is an ostial valve called the bulbo-ventricular valve. The conus arteriosus has a variable number of semilunar valves.

The ventral aorta delivers blood to the gills where it is oxygenated and flows, through the dorsal aorta, into the rest of the body. (In tetrapods, the ventral aorta has divided in two; one half forms the ascending aorta, while the other forms the pulmonary artery).

The circulatory systems of all vertebrates, are *closed*. Fish have the simplest circulatory system, consisting of only one circuit, with the blood being pumped through the capillaries of the gills and on to the capillaries of the body tissues. This is known as *single cycle* circulation.

In the adult fish, the four compartments are not arranged in a straight row but, instead form an S-shape with the latter two compartments lying above the former two. This relatively simpler pattern is found in cartilaginous fish and in the ray-finned fish. In teleosts, the conus arteriosus is very small and can more accurately be described as part of the aorta rather than of the heart proper. The conus arteriosus is not present in any amniotes, presumably having been absorbed into the ventricles over the course of evolution. Similarly, while the sinus venosus is present as a vestigial structure in some reptiles and birds, it is otherwise absorbed into the right atrium and is no longer distinguishable.

Swim Bladder

The swim bladder of a rudd

The swim bladder (or gas bladder) is an internal organ that contributes to the ability of a fish to control its buoyancy, and thus to stay at the current water depth, ascend, or descend without having to waste energy in swimming. The bladder is found only in the bony fishes. In the more primitive groups like some minnows, bichirs and lungfish, the bladder is open to the esophagus and doubles as a lung. It is often absent in fast swimming fishes such as the tuna and mackerel families. The condition of a bladder open to the esophagus is called physostome, the closed condition physoclist. In the latter, the gas content of the bladder is controlled through a rete mirabilis, a network of blood vessels effecting gas exchange between the bladder and the blood.

Weberian Apparatus

Fishes of the superorder Ostariophysi possess a structure called the Weberian apparatus, a modification which allow them to hear better. This ability which may well explain the marked success of otophysian fishes. The apparatus is made up of a set of bones known as *Weberian ossicles*, a chain of small bones that connect the auditory system to the swim bladder of fishes. The ossicles connect the gas bladder wall with Y-shaped lymph sinus that abuts the lymph-filled transverse canal joining the sacculi of the right and left ears. This allows the transmission of vibrations to the inner ear. A fully functioning Weberian apparatus consists of the swim bladder, the Weberian ossicles, a portion of the anterior vertebral column, and some muscles and ligaments.

Reproductive Organs

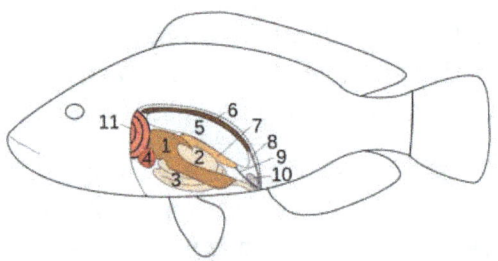

7 = testicles or ovaries

Fish reproductive organs include testes and ovaries. In most species, gonads are paired organs of similar size, which can be partially or totally fused. There may also be a range of secondary organs that increase reproductive fitness. The genital papilla is a small, fleshy tube behind the anus in some fishes, from which the sperm or eggs are released; the sex of a fish often can be determined by the shape of its papilla.

Testes

Most male fish have two testes of similar size. In the case of sharks, the testis on the right side is usually larger. The primitive jawless fish have only a single testis, located in the midline of the body, although even this forms from the fusion of paired structures in the embryo.

Under a tough membranous shell, the tunica albuginea, the testis of some teleost fish, contains very fine coiled tubes called seminiferous tubules. The tubules are lined with a layer of cells (germ cells) that from puberty into old age, develop into sperm cells (also known as spermatozoa or male gametes). The developing sperm travel through the seminiferous tubules to the rete testis located in the mediastinum testis, to the efferent ducts, and then to the epididymis where newly created sperm cells mature. The sperm move into the vas deferens, and are eventually expelled through the urethra and out of the urethral orifice through muscular contractions.

However, most fish do not possess seminiferous tubules. Instead, the sperm are produced in spherical structures called *sperm ampullae*. These are seasonal structures, releasing their contents during the breeding season, and then being reabsorbed by the body. Before the next breeding season, new sperm ampullae begin to form and ripen. The ampullae are otherwise essentially identical to the seminiferous tubules in higher vertebrates, including the same range of cell types.

In terms of spermatogonia distribution, the structure of teleosts testes has two types: in the most common, spermatogonia occur all along the seminiferous tubules, while in Atherinomorph fish they are confined to the distal portion of these structures. Fish can present cystic or semi-cystic spermatogenesis in relation to the release phase of germ cells in cysts to the seminiferous tubules lumen.

Ovaries

Many of the features found in ovaries are common to all vertebrates, including the presence of follicular cells and tunica albuginea There may be hundreds or even millions of fertile eggs present in the ovary of a fish at any given time. Fresh eggs may be developing from the germinal epithelium throughout life. Corpora lutea are found only in mammals, and in some elasmobranch fish; in other species, the remnants of the follicle are quickly resorbed by the ovary. The ovary of teleosts is often contains a hollow, lymph-filled space which opens into the oviduct, and into which the eggs are shed. Most normal female fish have two ovaries. In some elasmobranchs, only the right ovary develops fully. In the primitive jawless fish, and some teleosts, there is only one ovary, formed by the fusion of the paired organs in the embryo.

Fish ovaries may be of three types: gymnovarian, secondary gymnovarian or cystovarian. In the first type, the oocytes are released directly into the coelomic cavity and then enter the ostium, then through the oviduct and are eliminated. Secondary gymnovarian ovaries shed ova into the coelom from which they go directly into the oviduct. In the third type, the oocytes are conveyed to the exterior through the oviduct. Gymnovaries are the primitive condition found in lungfish, sturgeon, and bowfin. Cystovaries characterize most teleosts, where the ovary lumen has continuity with the oviduct. Secondary gymnovaries are found in salmonids and a few other teleosts.

Nervous System

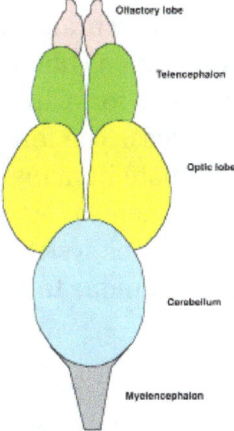

Dorsal view of the brain of the rainbow trout

Central Nervous System

Fish typically have quite small brains relative to body size compared with other vertebrates, typically one-fifteenth the brain mass of a similarly sized bird or mammal. However, some fish have relatively large brains, most notably mormyrids and sharks, which have brains about as massive relative to body weight as birds and marsupials.

Fish brains are divided into several regions. At the front are the olfactory lobes, a pair of structures that receive and process signals from the nostrils via the two olfactory nerves. Similar to the way humans smell chemicals in the air, fish smell chemicals in the water by tasting them. The olfactory lobes are very large in fish that hunt primarily by smell, such as hagfish, sharks, and catfish. Behind the olfactory lobes is the two-lobed telencephalon, the structural equivalent to the cerebrum in higher vertebrates. In fish the telencephalon is concerned mostly with olfaction. Together these structures form the forebrain.

The forebrain is connected to the midbrain via the diencephalon (in the diagram, this structure is below the optic lobes and consequently not visible). The diencephalon performs functions associated with hormones and homeostasis. The pineal body lies just above the diencephalon. This structure detects light, maintains circadian rhythms, and controls color changes. The midbrain or mesencephalon contains the two optic lobes. These are very large in species that hunt by sight, such as rainbow trout and cichlids.

The hindbrain or metencephalon is particularly involved in swimming and balance. The cerebellum is a single-lobed structure that is typically the biggest part of the brain. Hagfish and lampreys have relatively small cerebellae, while the mormyrid cerebellum is massive and apparently involved in their electrical sense.

The brain stem or myelencephalon is the brain's posterior. As well as controlling some muscles and body organs, in bony fish at least, the brain stem governs respiration and osmoregulation.

Vertebrates are the only chordate group to exhibit a proper brain. A slight swelling of the anterior end of the dorsal nerve cord is found in the lancelet, though it lacks the eyes and other complex sense organs comparable to those of vertebrates. Other chordates do not show any trends towards cephalisation. The central nervous system is based on a hollow nerve tube running along the length of the animal, from which the peripheral nervous system branches out to innervate the various systems. The front end of the nerve tube is expanded by a thickening of the walls and expansion of the central canal of spinal cord into three primary brain vesicles: The prosencephalon (forebrain), mesencephalon (midbrain) and rhombencephalon (hindbrain), further differentiated in the various vertebrate groups. Two laterally placed eyes form around outgrows from the midbrain, except in hagfish, though this may be a secondary loss. The forebrain is well developed and subdivided in most tetrapods, while the midbrain dominate in many fish and some salamanders. Vesicles of the forebrain are usually paired, giving rise to hemispheres like the cerebral hemispheres in mammals. The resulting anatomy of the central nervous system, with a single, hollow ventral nerve cord topped by a series of (often paired) vesicles is unique to vertebrates.

Cerebellum

The circuits in the cerebellum are similar across all classes of vertebrates, including fish, reptiles, birds, and mammals. There is also an analogous brain structure in cephalopods with well-developed brains, such as octopuses. This has been taken as evidence that the cerebellum performs functions important to all animal species with a brain.

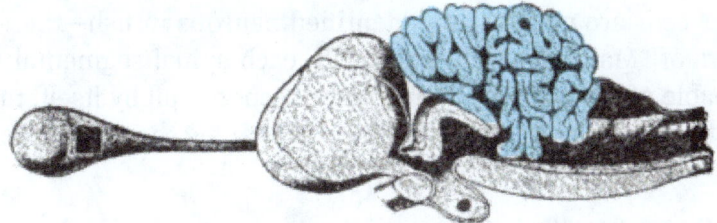

Cross-section of the brain of a porbeagle shark, with the cerebellum highlighted

There is considerable variation in the size and shape of the cerebellum in different vertebrate species. In amphibians, lampreys, and hagfish, the cerebellum is little developed; in the latter two groups, it is barely distinguishable from the brain-stem. Although the spinocerebellum is present in these groups, the primary structures are small paired nuclei corresponding to the vestibulocerebellum.

The cerebellum of cartilaginous and bony fishes is extraordinarily large and complex. In at least one important respect, it differs in internal structure from the mammalian cerebellum: The fish cerebellum does not contain discrete deep cerebellar nuclei. Instead, the primary targets of Purkinje cells are a distinct type of cell distributed across the cerebellar cortex, a type not seen in mammals. In mormyrid fish (a family of weakly electrosensitive freshwater fish), the cerebellum is considerably larger than the rest of the brain put together. The largest part of it is a special structure called the *valvula*, which has an unusually regular architecture and receives much of its input from the electrosensory system.

Most species of fish and amphibians possess a lateral line system that senses pressure waves in water. One of the brain areas that receives primary input from the lateral line organ, the medial octavolateral nucleus, has a cerebellum-like structure, with granule cells and parallel fibers. In electrosensitive fish, the input from the electrosensory system goes to the dorsal octavolateral nucleus, which also has a cerebellum-like structure. In ray-finned fishes (by far the largest group), the optic tectum has a layer — the marginal layer — that is cerebellum-like.

Identified Neurons

A neuron is called *identified* if it has properties that distinguish it from every other neuron in the same animal—properties such as location, neurotransmitter, gene expression pattern, and connectivity—and if every individual organism belonging to the same species has one and only one neuron with the same set of properties. In vertebrate nervous systems very few neurons are "identified" in this sense—in humans, there are believed to be none—but in simpler nervous systems, some or all neurons may be thus unique.

In vertebrates, the best known identified neurons are the gigantic Mauthner cells of fish. Every fish has two Mauthner cells, located in the bottom part of the brainstem, one on the left side and one on the right. Each Mauthner cell has an axon that crosses over, innervating neurons at the same brain level and then travelling down through the spinal cord, making numerous connections as it goes. The synapses generated by a Mauthner cell are so powerful that a single action potential gives rise to a major behavioral response: within milliseconds the fish curves its body into a C-shape, then straightens, thereby propelling itself rapidly forward. Functionally this is a fast escape response, triggered most easily by a strong sound wave or pressure wave impinging on the lateral line organ of the fish. Mauthner cells are not the only identified neurons in fish—there are about 20 more types, including pairs of "Mauthner cell analogs" in each spinal segmental nucleus. Although a Mauthner cell is capable of bringing about an escape response all by itself, in the context of ordinary behavior other types of cells usually contribute to shaping the amplitude and direction of the response.

Mauthner cells have been described as command neurons. A command neuron is a special type of identified neuron, defined as a neuron that is capable of driving a specific behavior all by itself. Such neurons appear most commonly in the fast escape systems of various species—the squid giant axon and squid giant synapse, used for pioneering experiments in neurophysiology because of their enormous size, both participate in the fast escape circuit of the squid. The concept of a command neuron has, however, become controversial, because of studies showing that some neurons that initially appeared to fit the description were really only capable of evoking a response in a limited set of circumstances.

Immune system

Immune organs vary by type of fish. In the jawless fish (lampreys and hagfish), true lymphoid organs are absent. These fish rely on regions of lymphoid tissue within other organs to produce immune cells. For example, erythrocytes, macrophages and plasma cells are produced in the anterior kidney (or pronephros) and some areas of the gut (where granulocytes mature.) They resemble primitive bone marrow in hagfish. Cartilaginous fish (sharks and rays) have a more advanced immune system. They have three specialized organs that are unique to chondrichthyes; the epigonal organs (lymphoid tissue similar to mammalian bone) that surround the gonads, the Leydig's organ within the walls of their esophagus, and a spiral valve in their intestine. These organs house typical immune cells (granulocytes, lymphocytes and plasma cells). They also possess an identifiable thymus and a well-developed spleen (their most important immune organ) where various lymphocytes, plasma cells and macrophages develop and are stored. Chondrostean fish (sturgeons, paddlefish and bichirs) possess a major site for the production of granulocytes within a mass that is associated with the meninges (membranes surrounding the central nervous system.) Their heart is frequently covered with tissue that contains lymphocytes, reticular cells and a small number of macrophages. The chondrostean kidney is an important hemopoietic organ; where erythrocytes, granulocytes, lymphocytes and macrophages develop.

Like chondrostean fish, the major immune tissues of bony fish (or teleostei) include the kidney (especially the anterior kidney), which houses many different immune cells. In addition, teleost fish possess a thymus, spleen and scattered immune areas within mucosal tissues (e.g. in the skin, gills, gut and gonads). Much like the mammalian immune system, teleost erythrocytes, neutrophils and granulocytes are believed to reside in the spleen whereas lymphocytes are the major cell type found in the thymus. In 2006, a lymphatic system similar to that in mammals was described in one species of teleost fish, the zebrafish. Although not confirmed as yet, this system presumably will be where naive (unstimulated) T cells accumulate while waiting to encounter an antigen.

References

- Kuraku; Hoshiyama, D; Katoh, K; Suga, H; Miyata, T; et al. (December 1999). "Monophyly of Lampreys and Hagfishes Supported by Nuclear DNA–Coded Genes". Journal of Molecular Evolution. 49 (6): 729–35. PMID 10594174. doi:10.1007/PL00006595

- Prosser, C. Ladd (1991). Comparative Animal Physiology, Environmental and Metabolic Animal Physiology (4th ed.). Hoboken, NJ: Wiley-Liss. pp. 1–12. ISBN 0-471-85767-X

- Guillaume, Jean; Praxis Publishing; Sadasivam Kaushik; Pierre Bergot; Robert Metailler (2001). Nutrition and Feeding of Fish and Crustaceans. Springer. p. 31. ISBN 1-85233-241-7. Retrieved 2009-01-09

- Nicholls, Henry (10 September 2009). "Mouth to Mouth". Nature. 461 (7261): 164–166. PMID 19741680. doi:10.1038/461164a

- Briggs, John C. (2005). "The biogeography of otophysian fishes (Ostariophysi: Otophysi): a new appraisal" (PDF). Journal of Biogeography. 32 (2): 287–294. doi:10.1111/j.1365-2699.2004.01170.x

- Liem, Karel F.; Warren Franklin Walker (2001). Functional anatomy of the vertebrates: an evolutionary perspective. Harcourt College Publishers. p. 277. ISBN 978-0-03-022369-3

- Alberts, Bruce; Johnson, Alexander; Lewis, Julian; Raff, Martin; Roberts, Keith; Walter, Peter (2002). Molecular Biology of the Cell (4th ed.). New York: Garland Science. Retrieved 2015-03-23

- Bell CC, Han V, Sawtell NB (2008). "Cerebellum-like structures and their implications for cerebellar function". Annu. Rev. Neurosci. 31: 1–24. PMID 18275284. doi:10.1146/annurev.neuro.30.051606.094225

- Calisher, CH (2007). "Taxonomy: what's in a name? Doesn't a rose by any other name smell as sweet?". Croatian Medical Journal. 48 (2): 268–270. PMC 2080517. PMID 17436393

- Romer, Alfred Sherwood; Parsons, Thomas S. (1977). The Vertebrate Body. Philadelphia, PA: Holt-Saunders International. pp. 129–145. ISBN 0-03-910284-X

- "Biodiversity: Mollusca". The Scottish Association for Marine Science. Archived from the original on 8 July 2006. Retrieved 2007-11-19

- Shi Z, Zhang Y, Meek J, Qiao J, Han VZ (2008). "The neuronal organization of a unique cerebellar specialization: the valvula cerebelli of a mormyrid fish". J. Comp. Neurol. 509 (5): 449–73. PMID 18537139. doi:10.1002/cne.21735

- Kim, Chang Bae; Moon, Seung Yeo; Gelder, Stuart R.; Kim, Won (September 1996). "Phylogenetic Relationships of Annelids, Molluscs, and Arthropods Evidenced from Molecules and Morphology". Journal of Molecular Evolution. New York: Springer. 43 (3): 207–215. PMID 8703086. doi:10.1007/PL00006079

- Gilbert, Scott F. (1994). Developmental Biology (4th ed.). Sunderland, Massachusetts: Sinauer Associates, Inc. p. 781. ISBN 0-87893-249-6

- Todaro, Antonio. "Gastrotricha: Overview". Gastrotricha: World Portal. University of Modena & Reggio Emilia. Retrieved 2008-01-26

- Koene, J. M. (2006). "Tales of two snails: sexual selection and sexual conflict in Lymnaea stagnalis and Helix aspersa". Integrative and Comparative Biology. 46 (4): 419–429. PMID 21672754. doi:10.1093/icb/icj040

2

Animal Anatomy: Cells and Tissues

Ameloblasts are the cells found during the development of teeth. Enteroendocrine cells are found in the gastrointestinal tract and the pancreas and help with the endocrine function. Cnidocyte, choanocyte, parietal cell, muscle tissue, pericyte and epithelium are the other types of cells and tissues explained in this section. This chapter provides a plethora of interdisciplinary topics for better comprehension of animal anatomy.

Ameloblast

Ameloblasts are cells present only during tooth development that deposit tooth enamel, which is the hard outermost layer of the tooth forming the surface of the crown.

Origin

Ameloblasts are derived from oral epithelium tissue of ectodermal origin. Their differentiation from preameloblasts (whose origin is from inner enamel epithelium) is a result of signaling from the ectomesenchymal cells of the dental papilla.

The ameloblasts will only become fully functional after the first layer of dentin (predentin) has been formed by odontoblasts. The cells are part of the reduced enamel epithelium after enamel maturation and then are subsequently lost during tooth eruption.

Life Cycle of Ameloblasts

The life cycle of ameloblasts consists of six stages :

1. Morphogenic stage
2. Organizing stage
3. Formative (secretory) stage (Tomes' processes appear)
4. Maturative stage
5. Protective stage
6. Desmolytic stage

The murine ALC cell line is of ameloblastic origin.

Function

Ameloblasts are cells which secrete the enamel proteins enamelin and amelogenin which will later mineralize to form enamel, the hardest substance in the human body. Ameloblasts control ionic and organic compositions of enamel. It is theorized that a circadian clock (24-hour) probably regulates enamel production on a daily cycle by the ameloblasts (similar to osteoblasts in production of bone tissue). Ameloblasts adjust their secretory and resorptive activities to maintain favorable conditions for biomineralization.

Structure

Each ameloblast is a columnar cell approximately 4 micrometers in diameter, 40 micrometers in length and is hexagonal in cross section. The secretory end of the ameloblast ends in a six-sided pyramid-like projection known as the Tomes' process. The angulation of the Tomes' process is significant in the orientation of enamel rods, the basic unit of tooth enamel.Distal terminal bars are junctional complexes that separate the Tomes' processes from ameloblast proper.

Pathophysiology

These cells are sensitive to their environment. One common example is illustrated by the neonatal line, a pronounced incremental line of Retzius found in the primary teeth and in the larger cusps of the permanent first molars, showing a disruption in enamel production when the person is born. High fevers in childhood are also an example of bodily stressors causing interruptions in enamel production.

Another possible example of this sensitivity (stress response pathway activation) may be the development of dental fluorosis after childhood exposure (between the ages of 2 to 8 years old) to excess fluoride, an elemental agent used to increase enamel hardness and prevent dental caries.

Myocyte

A myocyte (also known as a muscle cell) is the type of cell found in muscle tissue. Myocytes are long, tubular cells that develop from myoblasts to form muscles in a process known as myogenesis. There are various specialized forms of myocytes: cardiac, skeletal, and smooth muscle cells, with various properties. The striated cells of cardiac and skeletal muscles are referred to as muscle fibers. Cardiomyocytes are the muscle fibres that form the chambers of the heart, and have a single central nucleus. Skeletal muscle fibers help support and move the body and tend to have peripheral nuclei. Smooth muscle cells control involuntary movements such as the peristalsis contractions in the oesophagus and stomach.

Terminology

The unusual microstructure of muscle cells has led cell biologists to create specialized terminology. However, each term specific to muscle cells has a counterpart that is used in the terminology applied to other types of cells:

Muscle cell	Other organismal cells
sarcoplasm	cytoplasm
sarcoplasmic reticulum	smooth endoplasmic reticulum (SER)
sarcosome	mitochondrion
sarcolemma	cell membrane

The sarcoplasm is the cytoplasm of a muscle fiber. Most of the sarcoplasm is filled with myofibrils, which are long protein cords composed of myofilaments. The sarcoplasm is also composed of glycogen, a polysaccharide of glucose monomers, which provides energy to the cell with heightened exercise, and myoglobin, the red pigment that stores oxygen until needed for muscular activity.

There are three types of myofilaments:

- Thick filaments, composed of protein molecules called myosin. In striations of muscle bands, these are the dark filaments that make up the A band.

- Thin filaments are composed of protein molecules called actin. In striations of muscle bands, these are the light filaments that make up the I band.

- Elastic filaments are composed of titin, a large springy protein; these filaments anchor the thick filaments to the Z disc.

Together, these myofilaments work to produce a muscle contraction.

The sarcoplasmic reticulum, a specialized type of smooth endoplasmic reticulum, forms a network around each myofibril of the muscle fiber. This network is composed of groupings of two dilated end-sacs called terminal cisternae, and a single transverse tubule, or T tubule, which bores through the cell and emerge on the other side; together these three components form the triads that exist within the network of the sarcoplasmic reticulum, in which each T tubule has two terminal cisternae on each side of it. The sarcoplasmic reticulum serves as reservoir for calcium ions, so when an action potential spreads over the T tubule, it signals the sarcoplasmic reticulum to release calcium ions from the gated membrane channels to stimulate a muscle contraction.

The sarcolemma is the cell membrane of a striated muscle fiber and receives and conducts stimuli. At the end of each muscle fiber, the outer layer of the sarcolemma combines with tendon fibers. Within the muscle fiber pressed against the sarcolemma are multiple flattened nuclei; this multi-nuclear condition results from multiple myoblasts fusing to produce each muscle fiber, where each myoblast contributes one nucleus.

Structure

The cell membrane of a myocyte has several specialized regions, which may include the intercalated disk and the transverse tubular system. The cell membrane is covered by a lamina coat which is approximately 50 nm wide. The laminar coat is separable into two layers; the lamina densa and lamina lucida. In between these two layers can be several different types of ions, including calcium.

The cell membrane is anchored to the cell's cytoskeleton by anchor fibers that are approximately 10 nm wide. These are generally located at the Z lines so that they form grooves and transverse tubules emanate. In cardiac myocytes this forms a scalloped surface.

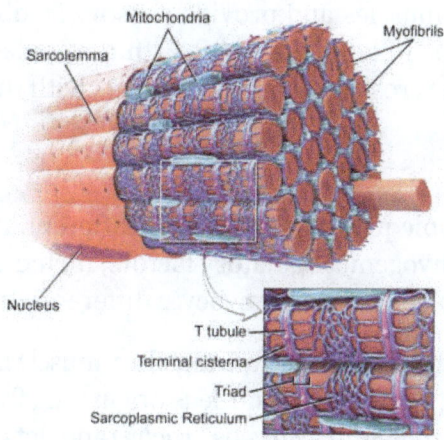

Skeletal Muscle Fiber

The cytoskeleton is what the rest of the cell builds off of and has two primary purposes; the first is to stabilize the topography of the intracellular components and the second is to help control the size and shape of the cell. While the first function is important for biochemical processes, the latter is crucial in defining the surface to volume ratio of the cell. This heavily influences the potential electrical properties of excitable cells. Additionally deviation from the standard shape and size of the cell can have negative prognostic impact.

Myofibrils

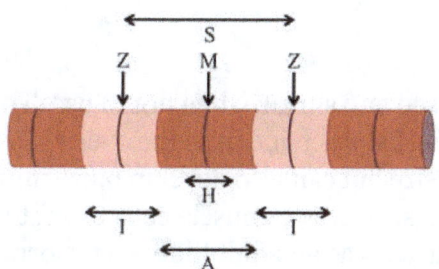

Portion of a myofibril, showing sarcomere structure:
S = Sarcomere, unit of muscle function   A = A-band, region of myosin   I = I-band, region of actin
H = H-zone, region of just myosin   Z = Z-line, sarcomere boundary   M = M-line, sarcomere center

Each muscle fiber contains myofibrils, which are very long chains of sarcomeres, the contractile units of the cell. A cell from the biceps brachii muscle may contain 100,000 sarcomeres. The myofibrils of smooth muscle cells are not arranged into sarcomeres. The sarcomeres are composed of thin and thick filaments. Thin filaments are made of actin and attach at Z lines which help them line up correctly with each other. Troponins are found at intervals along the thin filaments. Thick filaments are made of the elongated protein myosin. The sarcomere does not contain organelles or a nucleus. Sarcomeres are marked by Z lines which show the beginning and the end of a sarcomere. Individual myocytes are surrounded by endomysium.

Myocytes are bound together by perimysium into bundles called fascicles; the bundles are then grouped together to form muscle tissue, which is enclosed in a sheath of epimysium. The perimysium contains blood vessels and nerves which provide for the muscle fibers. Muscle spindles are distributed throughout the muscles and provide sensory feedback information to the central nervous system. Myosin is shaped like a long shaft with a rounded end pointed out towards the surface. This structure forms the cross bridge that connects with the thin filaments.

Formation From Myoblasts

A myoblast is a type of embryonic progenitor cell that differentiates to give rise to muscle cells. Differentiation is regulated by myogenic regulatory factors, including MyoD, Myf5, myogenin, and MRF4. GATA4 and GATA6 also play a role in myocyte differentiation.

Skeletal muscle fibers are made when myoblasts fuse together; muscle fibers therefore are cells with multiple nuclei, known as myonuclei, with each cell nucleus originating from a single myoblast. The fusion of myoblasts is specific to skeletal muscle (e.g., *biceps brachii*) and not cardiac muscle or smooth muscle.

Myoblasts in skeletal muscle that do not form muscle fibers dedifferentiate back into myosatellite cells. These satellite cells remain adjacent to a skeletal muscle fiber, situated between the sarcolemma and the basement membrane of the endomysium (the connective tissue investment that divides the muscle fascicles into individual fibers). To re-activate myogenesis, the satellite cells must be stimulated to differentiate into new fibers.

Myoblasts and their derivatives, including satellite cells, can now be generated in vitro through directed differentiation of pluripotent stem cells.

Kindlin-2 plays a role in developmental elongation during myogenesis.

Muscle Fiber Growth

Muscle fibers grow when exercised and shrink when not in use. This is due to the fact that exercise stimulates the increase in myofibrils which increase the overall size of muscle cells. Well exercised muscles can not only add more size but can also develop more mitochondria, myoglobin, glycogen and a higher density of capillaries. However muscle cells cannot divide to produce new cells, and as a result we have fewer muscle cells as an adult than a newborn.

Movement

When contracting, thin and thick filaments slide with respect to each other by using adenosine triphosphate. This pulls the Z discs closer together in a process called sliding filament mechanism. The contraction of all the sarcomeres results in the contraction of the whole muscle fiber. This contraction of the myocyte is triggered by the action potential over the cell membrane of the myocyte. The action potential uses transverse tubules to get from the surface to the interior of the myocyte, which is continuous within the cell membrane. Sarcoplasmic reticula are membranous bags that transverse tubules touch but remain separate from. These wrap themselves around each sarcomere and are filled with Ca^{2+}.

Excitation of a myocyte causes depolarization at its synapses, the neuromuscular junctions, which triggers action potential. With a singular neuromuscular junction, each muscle fiber receives input from just one somatic efferent neuron. Action potential in a somatic efferent neuron causes the release of the neurotransmitter acetylcholine.

When the acetylcholine is released it diffuses across the synapse and binds to a receptor on the sarcolemma, a term unique to muscle cells that refers to the cell membrane. This initiates an impulse that travels across the sarcolemma.

When the action potential reaches the sarcoplasmic reticulum it triggers the release of Ca^{2+} from the Ca^{2+} channels. The Ca^{2+} flows from the sarcoplasmic reticulum into the sarcomere with both of its filaments. This causes the filaments to start sliding and the sarcomeres to become shorter. This requires a large amount of ATP, as it is used in both the attachment and release of every myosin head. Very quickly Ca^{2+} is actively transported back into the sarcoplasmic reticulum, which blocks the interaction between the thin and thick filament. This in turn causes the muscle cell to relax.

Kinds of Contraction

There are four main different types of muscle contraction: twitch, treppe, tetanus and isometric/isotonic. Twitch contraction is the process previously described, in which a single stimulus signals for a single contraction. In twitch contraction the length of the contraction may vary depending on the size of the muscle cell. During treppe (or summation) muscles do not start at maximum efficiency, instead they achieve increased strength of contraction due to repeated stimuli. Tetanus involves a sustained contraction of muscles due to a series of rapid stimuli, which can continue until the muscles fatigue. Isometric are skeletal muscle contractions that do not cause movement of the muscle. However isotonic are skeletal muscles contractions that do cause movement.

Specialized cardiomyocytes located in the sinoatrial node are responsible for generating the electrical impulses that control the heart rate.

Fiber Typing

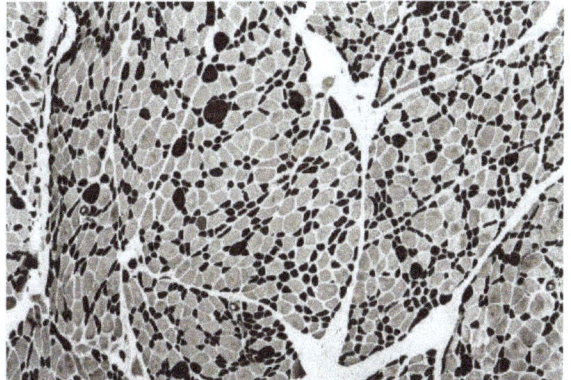

ATPase staining of a muscle cross section. Type II fibers are dark, due to the alkaline pH of the preparation. In this example, the size of the type II fibers is considerably less than the type I fibers due to denervation atrophy.

There are numerous methods employed for fiber-typing, and confusion between the methods is common among non-experts. Two commonly confused methods are histochemical staining for myosin ATPase activity and immunohistochemical staining for Myosin heavy chain (MHC) type. Myosin ATPase activity is commonly—and correctly—referred to as simply "fiber type", and results from the direct assaying of ATPase activity under various conditions (e.g. pH). Myosin heavy chain staining is most accurately referred to as "MHC fiber type", e.g. "MHC IIa fibers", and results from determination of different MHC isoforms. These methods are closely related physiologically, as the MHC type is the primary determinant of ATPase activity. However, that neither of these typing methods is directly metabolic in nature; they do not directly address oxidative or glycolytic capacity of the fiber. When "type I" or "type II" fibers are referred to generically, this most accurately refers to the sum of numerical fiber types (I vs. II) as assessed by myosin ATPase activity staining (e.g. "type II" fibers refers to type IIA + type IIAX + type IIXA... etc.).

Below is a table showing the relationship between these two methods, limited to fiber types found in humans. The sub-type capitalization used in fiber typing vs. MHC typing, and that some ATPase types actually contain multiple MHC types. Also, a subtype B or b is not expressed in humans by either method. Early researchers believed humans to express a MHC IIb, which led to the ATPase classification of IIB. However, later research showed that the human MHC IIb was in fact IIx, indicating that the IIB is better named IIX. IIb is expressed in other mammals, so is still accurately seen (along with IIB) in the literature. Non human fiber types include true IIb fibers, IIc, IId, etc.

ATPase Vs. MHC Fiber Types	
ATPase type	**MHC Heavy Chain(s)**
Type I	MHC Iβ
Type IC	MHC Iβ > MHC IIa
Type IIC	MHC IIa > MHC Iβ
Type IIA	MHC IIa
Type IIAX	MHC IIa > MHC IIx
Type IIXA	MHC IIx > MHC IIa
Type IIx	MHC IIx

Further fiber typing methods are less formally delineated, and exist on more of a spectrum. They tend to be focused more on metabolic and functional capacities (i.e., oxidative vs. glycolytic, fast vs. slow contraction time). As noted above, fiber typing by ATPase or MHC does not directly measure or dictate these parameters. However, many of the various methods are mechanistically linked, while others are correlated *in vivo*. For instance, ATPase fiber type is related to contraction speed, because high ATPase activity allows faster crossbridge cycling. While ATPase activity is only one component of contraction speed, type I fibers are "slow", in part, because they have low speeds of ATPase activity in comparison to type II fibers. However, measuring contraction speed is not the same as ATPase fiber typing.

Because of these types of relationships, Type I and Type II fibers have relatively distinct metabolic, contractile, and motor-unit properties. The table below differentiates these types of properties. However, it should be noted that these types of properties—while they are partly dependent on the properties of individual fibers—tend to be relevant and measured at the level of the motor unit, rather than individual fiber.

Various Properties of Different Fiber Types			
Properties	Type I fibers	Type IIA fibers	Type IIX fibers
Motor Unit Type	Slow Oxidative (SO)	Fast Oxidative/Glycolytic (FOG)	Fast Glycolytic (FG)
Twitch Speed	Slow	Fast	Fast
Twitch Force	Small	Medium	Large
Resistance to fatigue	High	High	Low
Glycogen Content	Low	High	High
Capillary Supply	Rich	Rich	Poor
Myoglobin	High	High	Low
Red Color	Dark	Dark	Pale
Mitochondrial density	High	High	Low
Capillary density	High	Intermediate	Low
Oxidative Enzyme Capacity	High	Intermediate-high	Low
Z-Line Width	Intermediate	Wide	Narrow
Alkaline ATPase Activity	Low	High	High
Acidic ATPase Activity	High	Medium-high	Low

Fiber Color

Traditionally, fibers were categorized depending on their varying color, which is a reflection of myoglobin content. Type I fibers appear red due to the high levels of myoglobin. Red muscle fibers tend to have more mitochondria and greater local capillary density. These fibers are more suited for endurance and are slow to fatigue because they use oxidative metabolism to generate ATP (adenosine triphosphate). Less oxidative type II fibers are white due to relatively low myoglobin and a reliance on glycolytic enzymes.

Twitch Speed

Fibers can also be classified on their twitch capabilities, into fast and slow twitch. These traits largely, but not completely, overlap the classifications based on color, ATPase, or MHC.

Some authors define a fast twitch fiber as one in which the myosin can split ATP very quickly. These mainly include the ATPase type II and MHC type II fibers However, fast twitch fibers also demonstrate a higher capability for electrochemical transmission of action potentials and a rapid level of calcium release and uptake by the sarcoplasmic reticulum. The fast twitch fibers rely on a well-developed, short term, glycolytic system for energy transfer and can contract and develop tension at 2–3 times the rate of slow twitch fibers. Fast twitch muscles are much better at generating short bursts of strength or speed than slow muscles, and so fatigue more quickly.

The slow twitch fibers generate energy for ATP re-synthesis by means of a long term system of aerobic energy transfer. These mainly include the ATPase type I and MHC type I fibers. They tend to have a low activity level of ATPase, a slower speed of contraction with a less well developed glycolytic capacity. They contain high mitochondrial volumes, and the high levels of myoglobin that give them a red pigmentation. They have been demonstrated to have high concentration of mitochondrial enzymes, thus they are fatigue resistant. Slow twitch muscles fire more slowly than fast twitch fibers, but are able to contract for a longer time before fatiguing.

Type Distribution

Individual muscles tend to be a mixture of various fiber types, but their proportions vary depending on the actions of that muscle and the species. For instance, in humans, the quadriceps muscles contain ~52% type I fibers, while the soleus is ~80% type I. The orbicularis oculi muscle of the eye is only ~15% type I. Motor units within the muscle, however, have minimal variation between the fibers of that unit. It is this fact that makes the size principal of motor unit recruitment viable.

The total number of skeletal muscle fibers has traditionally been thought not to change. It is believed there are no sex or age differences in fiber distribution, however, relative fiber types vary considerably from muscle to muscle and person to person. Sedentary men and women (as well as young children) have 45% type II and 55% type I fibers. People at the higher end of any sport tend to demonstrate patterns of fiber distribution e.g. endurance athletes show a higher level of type I fibers. Sprint athletes, on the other hand, require large numbers of type IIX fibers. Middle distance event athletes show approximately equal distribution of the two types. This is also often the case for power athletes such as throwers and jumpers. It has been suggested that various types of exercise can induce changes in the fibers of a skeletal muscle. It is thought that if you perform endurance type events for a sustained period of time, some of the type IIX fibers transform into type IIA fibers. However, there is no consensus on the subject. It may well be that the type IIX fibers show enhancements of the oxidative capacity after high intensity endurance training which brings them to a level at which they are able to perform oxidative metabolism as effectively as slow twitch fibers of untrained subjects. This would be brought about by an increase in mitochondrial size and number and the associated related changes not a change in fiber type.

Choanocyte

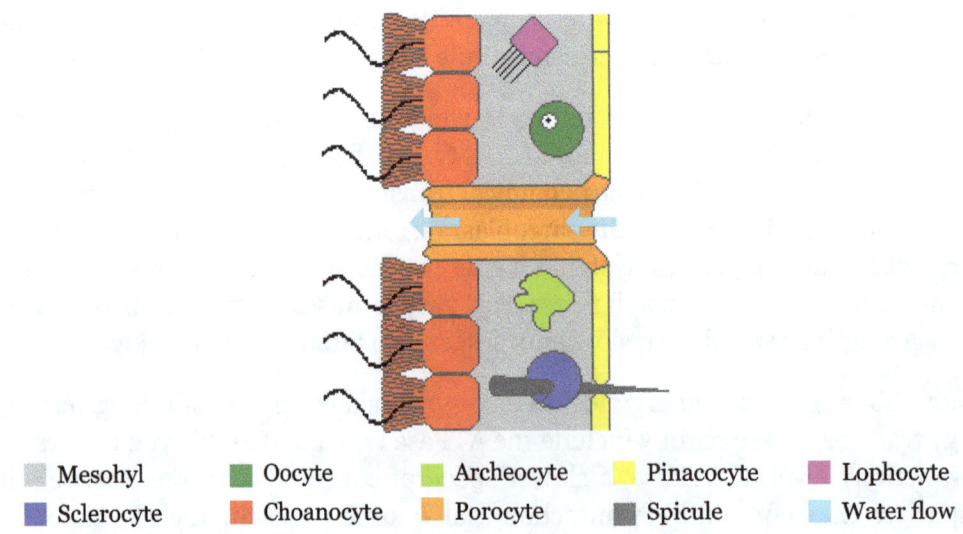

■ Mesohyl	■ Oocyte	■ Archeocyte	■ Pinacocyte	■ Lophocyte
■ Sclerocyte	■ Choanocyte	■ Porocyte	■ Spicule	■ Water flow

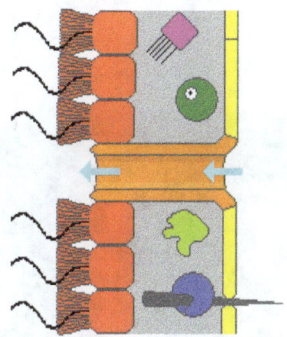

Main cell types of Porifera

Choanocytes (also known as "collar cells") are cells that line the interior of asconoid, syconoid and leuconoid body type sponges that contain a central flagellum, or *undulipodia*, surrounded by a collar of microvilli which are connected by a thin membrane. They make up Choanoderm, a type of cell layer found in sponges. The cell has the closest resemblance to the choanoflagellates which are the closest related single celled protists to the animal kingdom (metazoans). The flagellae beat regularly, creating a water flow across the microvilli which can then filter nutrients from the water taken from the collar of the sponge. Food particles are then phagocytosed by the cell.

Location

Choanocytes are found dotting the surface of the spongocoel in asconoid sponges and the radial canals in syconoid sponges, but they comprise entirely the chambers in leuconoid sponges.

Function

By cooperatively moving their flagella, choanocytes generate a flow of water through the sponges pores, into the spongocoel, and out through the osculum. This improves both respiratory and digestive functions for the sponge, pulling in oxygen and nutrients and allowing a rapid expulsion of carbon dioxide and other waste products. Although all cells in a sponge are capable of living on their own, choanocytes carry out most of the sponge's ingestion, passing digested materials to the amoebocytes for delivery to other cells.

Choanocytes can also turn into spermatocytes when needed for sexual reproduction, due to the lack of reproductive organs in sponges (amoebocytes become the oocytes).

Evolutionary Significance

Choanocytes bear more than a passing resemblance to Choanoflagellates, and demonstrate key steps in the evolution of animals. Scientist Nicole King helped to establish the distinction. DNA sequencing indicates that multicellular choanoflagellates and poriferans are sister groups, both descended from the same eukaryotic clade. One can see modern choanoflagellates living in small colonies, illustrating the evolution of sponges by analogy. More complex animals, notably the cnidarians, possess cells whose structures are clearly derived from choanocytes, demonstrating their historical ties to phylum porifera.

Cnidocyte

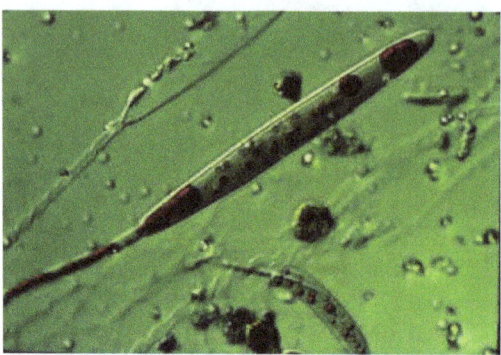

Nomarski micrograph of a ruthenium red-stained nematocyst from *Aiptasia pallida,* the pale anemone. The red dye stains the polyanionic venom proteins found inside the capsule of this partially discharged nematocyst.

A cnidocyte (also known as a cnidoblast or nematocyte) is an explosive cell containing one giant secretory organelle or *cnida* (plural *cnidae*) that defines the phylum Cnidaria (corals, sea anemones, hydrae, jellyfish, etc.). Cnidae are used for prey capture and defense from predators. Despite being morphologically simple, lacking a skeleton and many species being sessile, cnidarians prey on fish and crustaceans. A cnidocyte fires a structure that contains the toxin, from a characteristic subcellular organelle called a cnidocyst (also known as a cnida or nematocyst). This is responsible for the stings delivered by a cnidarian.

Structure and Function

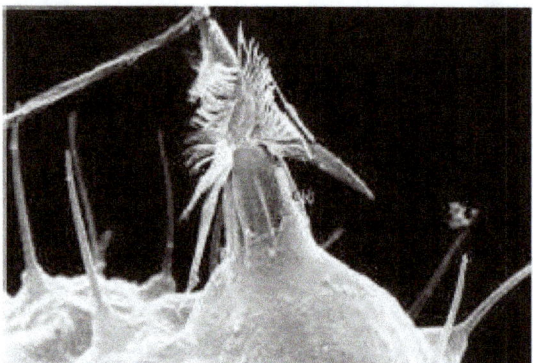

A discharged nematocyst seen under a scanning electron microscope

Each cnidocyte contains an organelle called a cnida or cnidocyst (e.g. nematocyst, ptychocyst or spirocyst), which comprises a bulb-shape capsule containing a coiled hollow tubule structure attached to it. The immature cnidocyte is referred to as a cnidoblast. The externally oriented side of the cell also has a hair-like trigger called a cnidocil. Cnidocil is a mechano and chemo-receptor. When the trigger is activated, the tubule shaft of the cnidocyst is ejected and in the case of the penetrant nematocyst, the forcefully ejected tubule penetrates the target organism. This discharge takes no more than a few microseconds, and is able to reach accelerations of about 40,000 g. Recent research suggests the process to occur as fast as 700 nanoseconds, thus reaching an acceleration of up to 5,410,000 g. After penetration, the toxic content of the nematocyst is injected into the target organism, allowing the sessile cnidarian to devour it.

Discharge Mechanism

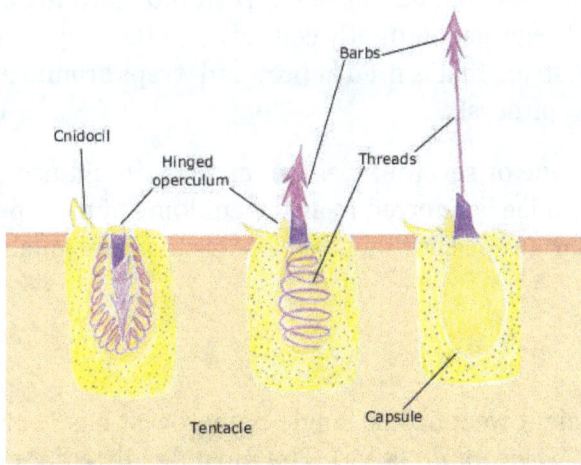

A diagram of the discharge mechanism of a nematocyst

Cnidae capsule stores a large concentration of calcium ions, which are released from the capsule into the cytoplasm of the *cnidocyte* when the trigger is activated. This causes a large concentration gradient of calcium across the cnidocyte plasma membrane. The resulting osmotic pressure causes a rapid influx of water into the cell. This increase in water volume in the cytoplasm forces the coiled cnidae tubule to eject rapidly. Prior to discharge the coiled cnidae tubule exists inside the cell in an "inside out" condition. The back pressure resulting from the influx of water into the cnidocyte together with the opening of the capsule tip structure or operculum, triggers the forceful eversion of the cnidae tubule causing it to right itself as it comes rushing out of the cell with enough force to impale a prey organism.

Prey Detection

Since cnidae are "single use" cells, and this costs a lot of energy, in order to regulate discharge, cnidocytes are connected as "batteries", containing several types of cnidocytes connected to supporting cells and neurons. The supporting cells contain chemosensors, which, together with the mechanoreceptor on the cnidocyte (cnidocil), allow only the right combination of stimuli to cause discharge, such as prey swimming, and chemicals found in prey cuticle or cuteous tissue. This prevents the cnidarian from stinging itself although sloughed off cnidae can be induced to fire independently.

Types of Cnidae

Over 30 types of cnidae are found in different cnidarians. They can be divided into the following groups:

1. Penetrant: The penetrant or stenotele is the largest and most complex nematocyst. When discharged, it pierces the skin or chitinous exoskeleton of the prey and injects the poisonous fluid, hypotoxin, that either paralyzes the victim or kills it.

2. Glutinant: a sticky surface used to stick to prey, referred to as ptychocysts and found on burrowing (tube) anemones, which help create the tube in which the animal lives

3. Volvent: The volvent or desmoneme is a small and pear-shaped nematocyst. It contains a short, thick, spineless, smooth and elastic thread tube forming a single loop and closed at the far end. When discharged, it tightly coils around the prey. They are the smallest nematocysts. A lasso-like string that is fired at prey and wraps around a cellular projection on the prey, referred to as spirocysts

Depending on the species, one or several types can appear simultaneously on the organism. The specific representation of cnidae is referred to as the cnidome of that species and may represent a dynamic aspect of the cnidarian species that is responsive to prey availability or the developmental stage of the organism.

Nematocyst Toxicity

Nematocysts are very efficient weapons. A single nematocyst has been shown to suffice in paralyzing a small arthropod (*Drosophila* larva). The most deadly cnidocytes (to humans, at least) are found on the body of a box jellyfish. One member of this family, the sea wasp, *Chironex fleckeri*, is "claimed to be the most venomous marine animal known," according to the Australian Institute of Marine Science. It can cause excruciating pain to humans, sometimes followed by death. Other cnidarians, such as the jellyfish *Cyanea capillata* (the "Lion's Mane" made famous by Sherlock Holmes) or the siphonophore *Physalia physalis* (Portuguese man o' war, "Bluebottle") can cause extremely painful and sometimes fatal stings. On the other hand, aggregating sea anemones may have the lowest sting intensity, perhaps due to the inability of the nematocysts to penetrate the skin, creating a feeling similar to touching sticky candies. Besides feeding and defense, sea anemone and coral colonies use cnidocytes to sting one another in order to defend or win space.

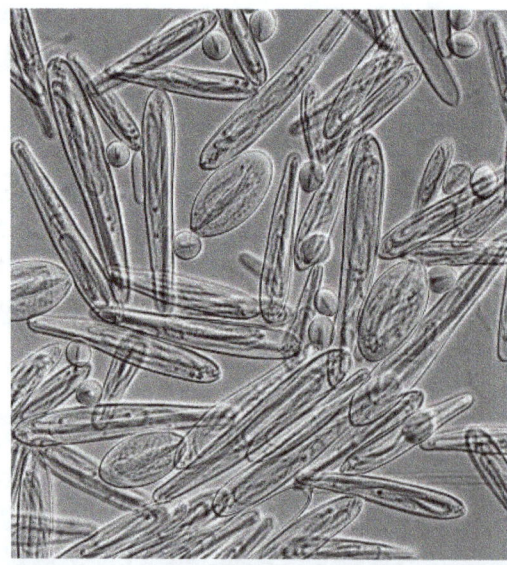

Nematocysts from *Chironex fleckeri* (400x magnification)

Venom from animals such as cnidarians, scorpions and spiders may be species-specific. A substance that is weakly toxic for humans or other mammals may be strongly toxic to the natural prey or predators of the venomous animal. Such specificity has been used to create new medicines and bioinsecticides.

Animals in the phylum Ctenophora ("Sea-Gooseberries" or "Comb Jellies") are transparent and jelly-like but have no nematocysts, and are harmless to humans.

Certain types of sea slugs, such as the nudibranch aeolids, are known to undergo kleptocnidae (in addition to kleptoplasty), whereby the organisms store nematocysts of digested prey at the tips of their cerata.

Enteroendocrine Cell

Enteroendocrine cells are specialized cells of the gastrointestinal tract and pancreas with endocrine function. They produce gastrointestinal hormones or peptides in response to various stimuli and release them into the bloodstream for systemic effect, diffuse them as local messengers, or transmit them to the enteric nervous system to activate nervous responses. Enteroendocrine cells of the intestine are the most numerous endocrine cells of the body. They constitute an enteric endocrine system as a subset of the endocrine system just as the enteric nervous system is a subset of the nervous system. In a sense they are known to act as chemoreceptors, initiating digestive actions and detecting harmful substances and initiating protective responses. Enteroendocrine cells are located in the stomach, in the intestine and in the pancreas.

Discovery

The very discovery of hormones occurred during studies of how the digestive system regulates its activities, as explained at *Secretin Discovery*.

Intestinal Enteroendocrine Cells

Intestinal enteroendocrine cells are not clustered together but spread as single cells throughout the intestinal tract.

Hormones secreted include somatostatin, motilin, cholecystokinin, neurotensin, vasoactive intestinal peptide, and enteroglucagon.

K Cell

K cells secrete gastric inhibitory peptide, an incretin, which also promotes triglyceride storage.

L Cell

L cells secrete glucagon-like peptide-1, an incretin, pancreatic peptide YY3-36, oxyntomodulin and glucagon-like peptide-2. L cells are primarily found in the ileum and large intestine (colon), but some are also found in the duodenum and jejunum.

I Cell

I cells secrete cholecystokinin (CCK), and are located in the duodenum and jejunum. They modulate bile secretion, exocrine pancreas secretion, and satiety.

G Cell

Stomach enteroendocrine cells, which release gastrin, and stimulate gastric acid secretion.

Enterochromaffin Cell

Enterochromaffin cells are enteroendocrine and neuroendocrine cells with a close similarity to adrenomedullary chromaffin cells secreting serotonin and histamine.

N Cell

Located in the jejunum, N cells release neurotensin, and control smooth muscle contraction.

S Cell

S cells secrete secretin from the duodenum and jejunum, and stimulate exocrine pancreatic secretion.

D Cell

also called Delta cells, secrete somatostatin

M Cell

secrete Motilin

Gastric Enteroendocrine Cells

Gastric enteroendocrine cells are found in the gastric glands, mostly at their base. The G cells secrete gastrin, post-ganglionic fibers of the vagus nerve can release gastrin-releasing peptide during parasympathetic stimulation to stimulate secretion. Enterochromaffin-like cells are enteroendocrine and neuroendocrine cells also known for their similarity to chromaffin cells secreting histamine, which stimulates G cells to secrete gastrin.

Other hormones produced include cholecystokinin, somatostatin, vasoactive intestinal peptide, substance P, alpha and gamma-endorphin.

Pancreatic Enteroendocrine Cells

Pancreatic enteroendocrine cells are located in the islets of Langerhans and produce most importantly the hormones insulin and glucagon. The autonomous nervous system strongly regulates their secretion, with parasympathetic stimulation stimulating insulin secretion and inhibiting glucagon secretion and sympathetic stimulation having opposite effect.

Other hormones produced include somatostatin, pancreatic polypeptide, amylin and ghrelin.

Pathology

Rare and slow growing carcinoid and non-carcinoid tumors develop from these cells. When a tumor arises it has the capacity to secrete large volumes of hormones.

Microfold Cell

Microfold cells (or M cells) are found in the gut-associated lymphoid tissue (GALT) of the Peyer's patches and in the mucosa-associated lymphoid tissue (MALT) of other parts of the gastrointestinal tract. These cells are known to initiate mucosal immunity responses on the apical membrane of the M cells and allow for transport of microbes and particles across the epithelial cell layer from the gut lumen to the lamina propria where interactions with immune cells can take place.

Unlike their neighbor cells, M cells have the unique ability to take up antigen from the lumen of the small intestine via endocytosis, phagocytosis, or transcytosis. Antigens are delivered to antigen presenting cells, such as dendritic cells, and lymphocytes (namely B cells). M cells express the protease cathepsin E, similar to other antigen presenting cells. This process takes place in a unique pocket-like structure on their basolateral side. Antigens are recognized via expression of cell surface receptors such as glycoprotein-2 (GP2) that detect and specifically bind to bacteria on bacteria. Cellular prion protein (PrP) is another example of a cell surface receptor on M cells.

M cells lack microvilli but, like other epithelial cells, they are characterized by strong cell junctions. This provides a physical barrier that constitutes an important line of defense between the gut contents and the immune system of the host. Despite the epithelial barrier, some antigens are able to infiltrate the M cell barrier and infect the nearby epithelial cells or enter the gut.

Morphology and Function

M cells are distinguished from other intestinal epithelial cells by their morphological differences. They are characterized by their short microvilli or lack of these protrusions on the cell surface. When they present microvilli, they are short, irregular, and present on the apical surface or pocket-like invagination on the basolateral surface of these cells. When they lack microvilli, they are characterized by their microfolds, and hence receive their commonly known name. These cells are far less abundant than enterocytes. These cells can also be identified by cytoskeletal and extracellular matrix components expressed at the edge of cells or on their cell surfaces, such as actin, villin, cytokeratin, and vimentin.

M cells do not secrete mucus or digestive enzymes, and have a thinner glycocalyx, which allows them to have easy access to the intestinal lumen for endocytosis of antigens. The main function of M cells is the selective endocytosis of antigens, and transporting them to intraepithelial macrophages and lymphocytes, which then migrate to lymph nodes where an immune response can be initiated

Pathology

M cells are exploited by several pathogenic gram-negative bacteria including *Shigella flexneri*, *Salmonella typhimurium*, and *Yersinia pseudotuberculosis*, as well as infectious prions, such as in Bovine spongiform encephalitis (Mad-cow disease), as a way of penetrating the intestinal epithelium. Exploitation as a virulence factor depends upon the pathogen's ability to bind to M cells and thus guarantee penetration in that manner, as M cells sample intestinal contents. EPEC containing plasmids with genes for EAF (*Escherichia coli* Adherence Factor) will adhere to M cells.

They are also exploited by viruses such as Polio and Reovirus for dissemination. CXCR4 tropic but not CCR5 tropic HIV has been noted to be able to bind to M cells and get transported across the epithelium by them.

Development

Factors promoting the differentiation of M cells have yet to be elucidated, but they are thought to develop in response to signals from immune cells found in developing Peyer's patches. B cells have been implicated in the developmental of M cells, since they are also localized in high numbers in the follicular-associated epithelium (FAE). FAE lacking B cell populations results in a decrease in the number of M cell lining the Peyer's patches. Similarly, a human lymphoma cell line is also known to undergo transition from adenocarcinoma cells to M cells.

Though many studies have shown various cell types directing the differentiation of M cells, new research characterizes the molecular pathways that guide M cell differentiation. More recently, through loss-of-function and rescue-phenotype studies, RANKLis shown to be a receptor activator of NF-κB ligand and play a role in differentiation of M cells. RANKL is expressed throughout the small intestine, facilitates uptake of pathogens such as Salmonella, and is the most critical factor M cell differentiation. Interestingly, microbes found on intestinal epithelium are known to direct M cell development. For example, the type III secretion system effector protein SopB activates the transition of M cells from enterocytes. M cells undergo the differentiation process for up to four days before reaching full maturation. Recent studies have suggested they arise distinctly from the lymphoid and myeloid lineages.

Pathogens can take advantage of cell differentiation pathways in order to invade host cells. This is done by inducing differentiation of enterocytes into M cell type in gut epithelium. In one case, the SopB effector protein mentioned above is secreted to trigger fast differentiation of enterocytes localized in the FAE by initiation of epithelial to mesenchymal transition in these cells. When SopB activates differentiation of enterocytes, it acts via the activation of the Wnt/b-catenin signaling pathway and triggers the RANKL and its receptor, implicated in regulating cell apoptosis.

Passive Immunity in Humans

M cells play a role in passive immunity, or the transfer of active humoral immunity during and post pregnancy. Infants rely on antibodies specific to their mother's intestinal antigens, which move from the mother's gut and enter the breast milk. These antibodies are able to move into the milk supply through the lymphatic system. Even though the mechanism of this transport is not fully understood, it is hypothesized that dendritic cells and macrophages play the role of transport vehicles. In females that are not lactating, when M cells recognize antigen in the gut, they stimulate production of many Immunoglobulin A (IgA) antibodies. These antibodies are released into the gut mucosa, salivary glands, and lymph nodes. However, in females that are lactating, M cells recognize antigen and IgA is directed from the gut to the mammary gland. IgA traveling from the gut to breast milk supply is controlled by hormones, chemokines, and cytokines. Thus, the mammary gland and breast milk have critical roles alongside M cells in mucosal immune system.

Parietal Cell

Parietal cells (also known as oxyntic or delomorphous cells), are the epithelial cells that secrete hydrochloric acid (HCl) and intrinsic factor. These cells are located in the gastric glands found in the lining of the fundus and in the body of the stomach. They contain an extensive secretory network (called canaliculi) from which the HCl is secreted by active transport into the stomach. The enzyme hydrogen potassium ATPase (H^+/K^+ ATPase) is unique to the parietal cells and transports the H^+ against a concentration gradient of about 3 million to 1, which is the steepest ion gradient formed in the human body. Parietal cells are primarily regulated via histamine, acetylcholine and gastrin signaling from both central and local modulators.

Secretion of Hydrochloric Acid

Synthesis

Hydrochloric acid is formed in the following manner:

- Hydrogen ions are formed from the dissociation of carbonic acid. Water is a very minor source of hydrogen ions in comparison to carbonic acid. Carbonic acid is formed from carbon dioxide and water by carbonic anhydrase.

- The bicarbonate ion (HCO_3^-) is exchanged for a chloride ion (Cl^-) on the basal side of the cell and the bicarbonate diffuses into the venous blood, leading to an alkaline tide phenomenon.

- Potassium (K^+) and chloride (Cl^-) ions diffuse into the canaliculi.

- Hydrogen ions are pumped out of the cell into the canaliculi in exchange for potassium ions, via the H^+/K^+ ATPase. These receptors are increased in number on luminal side by fusion of tubulovesicles during activation of parietal cells and removed during deactivation. This receptor maintains a million fold difference in proton concentration. ATP is provided by the numerous mitochondria.

As a result of the cellular export of hydrogen ions, the gastric lumen is maintained as a highly-acidic environment. The acidity aids in digestion of food by promoting the unfolding (or denaturing) of ingested proteins. As proteins unfold, the peptide bonds linking component amino acids are exposed. Gastric HCl simultaneously activates pepsinogen, an endopeptidase that advances the digestive process by breaking the now-exposed peptide bonds, a process known as proteolysis.

Regulation

Parietal cells secrete acid in response to three types of stimuli:

- Histamine, stimulates H_2 histamine receptors (most significant contribution).

- Acetylcholine, from parasympathetic activity via the vagus nerve and enteric nervous system, stimulating M_3 receptors.

- Gastrin, stimulating CCK2 receptors (least significant contribution, but also causes histamine secretion by local ECL cells)

Activation of histamine through H_2 receptor causes increases intracellular cAMP level while Ach through M_3 receptor and gastrin through CCK2 receptor increases intracellular calcium level. These receptors are present on basolateral side of membrane.

Increased cAMP level results in increased protein kinase A. Protein kinase A phosphorylates proteins involved in the transport of H^+/K^+ ATPase from the cytoplasm to the cell membrane. This causes resorption of K^+ ions and secretion of H^+ ions. The pH of the secreted fluid can fall by 0.8.

Gastrin primarily induces acid-secretion indirectly, increasing histamine synthesis in ECL cells,which in turn signal parietal cells via histamine release/H2 stimulation. Gastrin itself has no effect on the maximum histamine-stimulated gastric acid secretion.

The effect of histamine, acetylcholine and gastrin is synergistic, that is, effect of two simultaneously is more than additive of effect of the two individually. It helps in non-linear increase of secretion with stimuli physiologically.

Intrinsic Factor

Parietal cells also produce intrinsic factor. Intrinsic factor is required for the absorption of vitamin B_{12} in the diet. A long-term deficiency in vitamin B_{12} can lead to megaloblastic anemia, characterized by large fragile erythrocytes. Pernicious anaemia results from autoimmune destruction of gastric parietal cells, precluding the synthesis of intrinsic factor and, by extension, absorption of Vitamin B_{12}. Pernicious anemia also leads to megaloblastic anemia. Atrophic gastritis, particularly in the elderly, will cause an inability to absorb B_{12} and can lead to deficiencies such as decreased DNA synthesis and nucleotide metabolism in the bone marrow.

Canaliculus

A canaliculus is an adaptation found on gastric parietal cells. It is a deep infolding, or little channel, which serves to increase the surface area, e.g. for secretion. The parietal cell membrane is dynamic; the numbers of canaliculi rise and fall according to secretory need. This is accomplished by the fusion of canalicular precursors, or "tubulovesicles", with the membrane to increase surface area, and the reciprocal endocytosis of the canaliculi (reforming the tubulovesicles) to decrease it.

Diseases of Parietal Cells

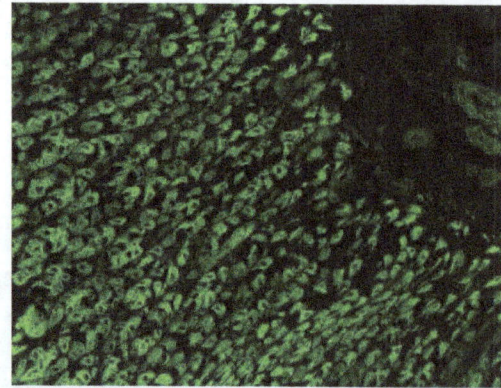

Immunofluorescence staining pattern of gastric parietal antibodies on a stomach section

- *Peptic ulcers* can result from over-acidity in the stomach. Antacids can be used to enhance the natural tolerance of the gastric lining. Antimuscarinic drugs such as pirenzepine or H_2 antihistamines can reduce acid secretion. Proton pump inhibitors are more potent at reducing gastric acid production since that is the final common pathway of all stimulation of acid production.

- In *pernicious anemia*, autoantibodies directed against parietal cells or intrinsic factor cause a reduction in vitamin B_{12} absorption. It can be treated with injections of replacement vitamin B_{12} (methylcobalamin, hydroxocobalamin or cyanocobalamin).

- *Achlorhydria* is another autoimmune disease of the parietal cells. The damaged parietal cells are unable to produce the required amount of gastric acid. This leads to an increase in gastric pH, impaired digestion of food and increased risk of gastroenteritis.

Pericyte

Pericytes are contractile cells that wrap around the endothelial cells that line the capillaries and venules throughout the body. Also known as Rouget cells or mural cells, pericytes are embedded in basement membrane where they communicate with endothelial cells of the body's smallest blood vessels by means of both direct physical contact and paracrine signaling. In the brain, pericytes help sustain the blood–brain barrier as well as several other homeostatic and hemostatic functions of the brain. These cells are also a key component of the *neurovascular unit*, which includes endothelial cells, astrocytes, and neurons. Pericytes regulate capillary blood flow, the clearance and phagocytosis of cellular debris, and the permeability of the blood–brain barrier. Pericytes stabilize and monitor the maturation of endothelial cells by means of direct communication between the cell membrane as well as through paracrine signaling. A deficiency of pericytes in the central nervous system can cause the blood–brain barrier to break down.

Morphology

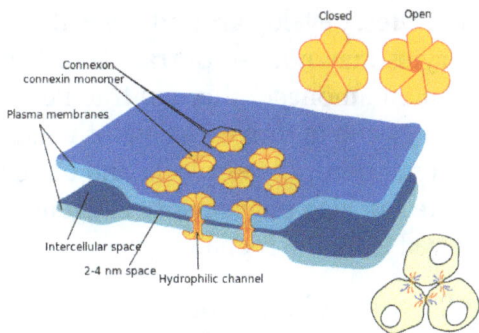

Gap cell junction created between two neighboring cells by connexin.

In the central nervous system, pericytes wrap around the endothelial cells that line the inside of the capillary. These two types of cells can be easily distinguished from one another based on the presence of the prominent round nucleus of the pericyte compared to the flat elongated nucleus of the endothelial cells. Pericytes also project finger-like extensions that wrap around the capillary wall, allowing the cells to regulate capillary blood flow.

Both pericytes and endothelial cells share a basement membrane where a variety of intercellular connections are made. Many types of integrin molecules facilitate communication between pericytes and endothelial cells separated by the basement membrane. Pericytes can also form direct connections with neighboring cells by forming peg and socket arrangements in which parts of the cells interlock, similar to the gears of a clock. At these interlocking sites, gap junctions can be formed which allow the pericytes and neighboring cells to exchange ions and other small molecules. Important molecules in these intercellular connections include N-cadherin, fibronectin, connexin and various integrins.

In some regions of the basement membrane, adhesion plaques composed of fibronectin can be found. These plaques facilitate the connection of the basement membrane to the cytoskeletal structure composed of actin, and the plasma membrane of the pericytes and endothelial cells.

Function

Skeletal Muscle Regeneration and Fat Formation

Pericytes in the skeletal striated muscle are of two distinct populations, each with its own role. The first pericyte subtype (Type-1) can differentiate into fat cells while the other (Type-2) into muscle cells. Type-1 characterized by negative expression for nestin (PDGFRβ+CD146+NG2-) and type-2 characterized by positive expression for nestin (PDGFRβ+CD146+NG2+). While both types are able to proliferate in response to glycerol or BaCl2-induced injury, type-1 pericytes give rise to adipogenic cells only in response to glycerol injection and type-2 become myogenic in response to both types of injury. The extent to which type-1 pericytes participate in fat accumulation is not known.

Angiogenesis and the Survival of Endothelial Cells

Pericytes are also associated with allowing endothelial cells to differentiate, multiply, form vascular branches (angiogenesis), survive apoptotic signals and travel throughout the body. Certain pericytes, known as microvascular pericytes, develop around the walls of capillaries and help to serve this function. Microvascular pericytes may not be contractile cells because they lack alpha-actin isoforms; structures that are common amongst other contractile cells. These cells communicate with endothelial cells via gap junctions and in turn cause endothelial cells to proliferate or be selectively inhibited. If this process did not occur, hyperplasia and abnormal vascular morphogenesis could occur. These types of pericyte can also phagocytose exogenous proteins. This suggests that the cell type might have been derived from microglia.

A lineage relationship to other cell types has been proposed, including smooth muscle cells, neural cells, NG2 glia, muscle fibers, adipocytes, as well as fibroblasts and other mesenchymal stem cells, however whether these cells differentiate into each other is an outstanding question in the field. Pericytes' regenerative capacity is affected by aging. Such versatility is conducive because they actively remodel blood vessels throughout the body and can thereby blend homogeneously with the local tissue environment.

Aside from creating and remodeling blood vessels in a viable fashion, pericytes have been found to protect endothelial cells from death via apoptosis or cytotoxic elements. It has been studied *in vivo* that pericytes release a hormone known as pericytic aminopeptidase N/pAPN that may help to promote angiogenesis. When this hormone was mixed with cerebral endothelial cells as well as astrocytes, the pericytes grouped into structures that resembled capillaries. Furthermore, if experimental group contained all of the following with the exception of pericytes, the endothelial cells would undergo apoptosis. That being said, it was concluded that pericytes must be present to assure the proper function of endothelial cells and astrocytes must be present to assure that both remain in contact. If not, then proper angiogenesis cannot occur. In addition, it has been found that pericytes contribute to the survival of endothelial cells because they secrete the protein Bcl-w during cellular crosstalk. Bcl-w is an instrumental protein in the pathway that enforces VEGF-A expression and discourages apoptosis. Although there is some speculation as to why VEGF is directly responsible for preventing apoptosis, it is believed to be responsible for modulating apoptotic signal transduction pathways and inhibiting activation of apoptosis inducing enzymes. Two biochemical mechanisms utilized by VEGF to accomplish such would be phosphorylation of extracellular regulatory kinase 1 (ERK-1) which sustains cell survival over time and inhibition of stress-activated protein kinase/c-jun-NH2 kinase which also promotes apoptosis.

Blood–brain Barrier

Pericytes play a crucial role in the formation and functionality of the selectively permeable space between the circulatory system and central nervous system. This space is known as the blood–brain barrier. This barrier is composed of endothelial cells and assures the protection and functionality of the brain and central nervous system. Although it had been theorized that astrocytes were crucial to the postnatal formation of this barrier, it has been found that pericytes are now largely responsible for this role. Pericytes are responsible for tight junction formation and vesicle trafficking amongst endothelial cells. Furthermore, they allow the formation of the blood–brain barrier by inhibiting the effects of CNS immune cells (which can damage the formation of the barrier) and by reducing the expression of molecules that increase vascular permeability.

Aside from blood–brain barrier formation, pericytes also play an active role in its functionality by controlling the flow within blood vessels and between blood vessels and the brain. In animal models with lower pericyte coverage, trafficking of molecules across endothelial cells occurs at a higher frequency, allowing proteins into the brain that would normally be excluded. Loss or dysfunction of pericytes is also theorized to contribute to neurodegenerative diseases such as Alzheimer's, Parkinson's and ALS (Lou Gehrig's Disease) through breakdown of the blood-brain barrier.

Blood Flow

Increasing evidence suggests that pericytes can regulate blood flow at the capillary level. For the retina, movies have been published showing that pericytes constrict capillaries when their membrane potential is altered to cause calcium influx, and in the brain it has been reported that neuronal activity increases local blood flow by inducing pericytes to dilate capillaries before upstream arteriole dilation occurs. This area is controversial, with a recent study claiming that pericytes do not express contractile proteins and are not capable of contraction in vivo, although the latter paper has been criticised for using a highly unconventional definition of pericyte which explicitly excludes contractile pericytes. It appears that different signaling pathways regulate the constriction of capillaries by pericytes and of arterioles by smooth muscle cells

Pericytes are important in maintaining circulation. In a study involving adult pericyte-deficient mice, cerebral blood flow was diminished with concurrent vascular regression due to loss of both endothelia and pericytes. Significantly greater hypoxia was reported in the hippocampus of pericyte-deficient mice as well as inflammation, and learning and memory impairment.

Pathologies

Because of their crucial role in maintaining and regulating endothelial cell structure and blood flow, abnormalities in pericyte function are seen in many pathologies. They may either be present in excess, leading to diseases such as hypertension and tumor formation, or in deficiency, leading to neurodegenerative diseases.

Hemangioma

The clinical phases of Hemangioma have physiological differences, correlated with immunophenotypic profiles by Takahashi et al. During the early proliferative phase (0–12 months) the tumors express proliferating cell nuclear antigen (pericytesna), vascular endothelial growth factor (VEGF), and type IV collagenase, the former two localized to both endothelium and pericytes, and the last to endothelium. The vascular markers CD 31, von Willebrand factor (vWF), and smooth muscle actin (pericyte marker) are present during the proliferating and involuting phases, but are lost after the lesion is fully involuted.

Hemangiopericytoma

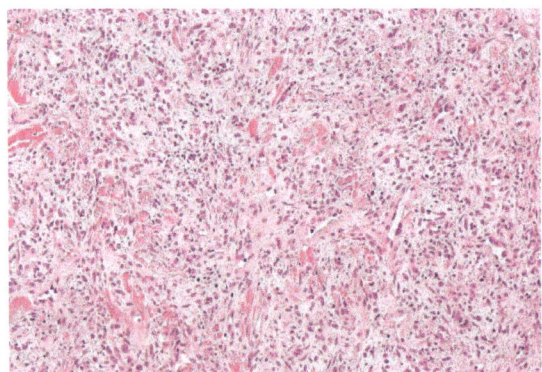

Image of a solitary fibrous tumour that is most likely a hemangiopericytoma. It surrounds a staghorn-shaped blood vessel, which results from the arrangement of pericytes around the vessel

Hemangiopericytoma is a rare vascular neoplasm, or abnormal growth, that may either be benign or malignant. In its malignant form, metastasis to the lungs, liver, brain, and extremities may occur. It most commonly manifests itself in the femur and proximal tibia as a bone sarcoma, and is usually found in older individuals, though cases have been found in children. Hemangiopericytoma is caused by the excessive layering of sheets of pericytes around improperly formed blood vessels. Diagnosis of this tumor is difficult because of the inability to distinguish pericytes from other types of cells using light microscopy. Treatment may involve surgical removal and radiation therapy, depending on the level of bone penetration and stage in the tumor's development.

Diabetic Retinopathy

The retina of diabetic individuals often exhibits loss of pericytes, and this loss is a characteristic factor of the early stages of diabetic retinopathy. Studies have found that pericytes are essential in diabetic individuals to protect the endothelial cells of retinal capillaries. With the loss of pericytes, microaneurysms form in the capillaries. In response, the retina either increases its vascular permeability, leading to swelling of the eye through a macular edema, or forms new vessels that permeate into the vitreous membrane of the eye. The end result is reduction or loss of vision. While it is unclear why pericytes are lost in diabetic patients, one hypothesis is that toxic sorbitol and advanced glycation end-products (AGE) accumulate in the pericytes. Because of the build-up of glucose, the polyol pathway increases its flux, and intracellular sorbitol and fructose accumulate. This leads to osmotic imbalance, which results in cellular damage. The presence of high glucose levels also leads to the buildup of AGE's, which also damage cells.

Neurodegenerative Diseases

Studies have found that pericyte loss in the adult and aging brain leads to the disruption of proper cerebral perfusion and maintenance of the blood–brain barrier, which causes neurodegeneration and neuroinflammation. The apoptosis of pericytes in the aging brain may be the result of a failure in communication between growth factors and receptors on pericytes. Platelet-derived growth factor B (PDGFB) is released from endothelial cells in brain vasculature and binds to the receptor PDGFRB on pericytes, initiating their proliferation and investment in the vasculature.

Immunohistochemical studies of human tissue from Alzheimer's disease and amyotrophic lateral sclerosis show pericyte loss and breakdown of the blood-brain barrier. Pericyte-deficient mouse models (which lack genes encoding steps in the PDGFB:PDGFRB signalling cascade) and have an Alzheimer's-causing mutation have exacerbated Alzheimer's-like pathology compared to mice with normal pericyte coverage and an Alzheimer's-causing mutation.

Stroke

In conditions of stroke pericytes constrict brain capillaries and then die, which may lead to a long-lasting decrease of blood flow and loss of blood–brain barrier function, increasing the death of nerve cells

Current Research

Endothelial and Pericyte Interactions

Endothelial cells and pericytes are interdependent, so failure of proper communication between the two cells can lead to numerous human pathologies.

There are several pathways of communication between the endothelial cells and pericytes. The first is transforming growth factor (TGF) signaling, which is mediated by endothelial cells. This is important for pericyte differentiation. Angiopoietin 1 and Tie-2 signaling is essential for maturation and stabilization of endothelial cells. Platelet-derived growth factor (PDGF) pathway signaling from endothelial cells recruits pericytes, so that pericytes can migrate to growing vessels. If this pathway is blocked, it leads to pericyte deficiency. Sphingosine-1-phosphate (S1P) signaling also aides in pericyte recruitment by communication through G protein-coupled receptors. S1P signals through GTPases that promote N-cadherin trafficking to endothelial membranes. This trafficking strengthens contacts with pericytes.

Communication between endothelial cells and pericytes is important. Inhibiting the PDGF pathway leads to pericyte deficiency. This causes endothelial hyperplasia, abnormal junctions, and diabetic retinotropy. A lack of pericytes also causes an upregulation of vascular endothelial growth factor (VEGF), leading to vascular leakage and hemorrhage. Also, angiopoietin 2 can act as an antagonist to Tie-2. This destabilizes the endothelial cells, which accounts for less endothelial cell and pericyte interaction. This can actually lead to the formation of tumors. Similar to the inhibition of the PDGF pathway, angiopoietin 2 reduces levels of pericytes, leading to diabetic retinopathy.

Scarring

After an injury in the central nervous system, scarring occurs to preserve the integrity of surrounding cells. Usually, astrocytes are associated with the scarring and are called glial scars. However, there is a stromal or nonglial component of the scarring, and lineage-tracking studies of pericytes following stroke revealed that they form the main component of the glial scar. Following traumatic brain injury, pericytes migrate out from the vasculature to the site of injury, differentiate into myofibroblasts and deposit extracellular matrix that forms the fibrotic component of the scar.

The scarring is highly compartmentalized. The pericytes form the core of the scar, while ependymal cells form a second layer around the core, followed by another layer of astrocytes that originated through self-duplication.

Inhibition of subtype A pericyte generation caused improper closing of spinal cord incisions, which supports the idea that pericytes are important for scarring.

Epithelium

Epithelium is one of the four basic types of animal tissue, along with connective tissue, muscle tissue and nervous tissue. Epithelial tissues line the cavities and surfaces of blood vessels and organs throughout the body.

There are three principal shapes of epithelial cell: squamous, columnar, and cuboidal. These can be arranged in a single layer of cells as simple epithelium, either squamous, columnar, cuboidal, pseudo-stratified columnar or in layers of two or more cells deep as stratified (layered), either squamous, columnar or cuboidal. All glands are made up of epithelial cells. Functions of epithelial cells include secretion, selective absorption, protection, transcellular transport, and sensing.

Epithelial layers contain no blood vessels, so they must receive nourishment via diffusion of substances from the underlying connective tissue, through the basement membrane. Cell junctions are well-employed in epithelial tissues.

Classification

In general, epithelial tissues are classified by the number of their layers and by the shape and function of the cells.

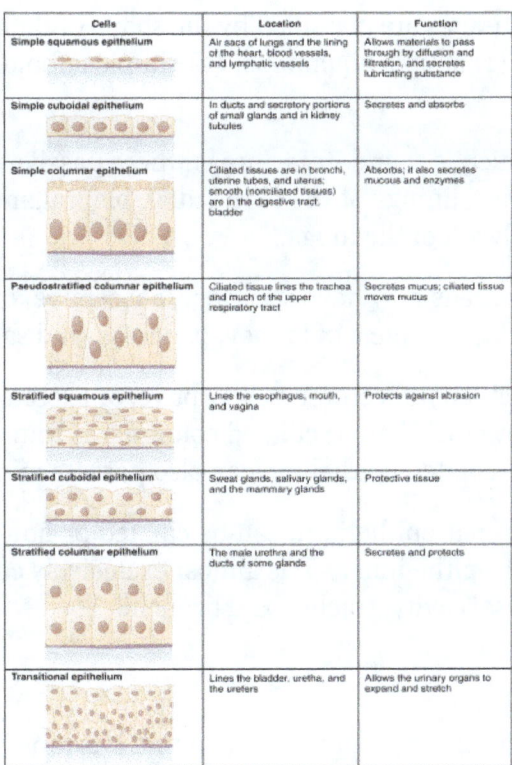

Cells	Location	Function
Simple squamous epithelium	Air sacs of lungs and the lining of the heart, blood vessels, and lymphatic vessels	Allows materials to pass through by diffusion and filtration, and secretes lubricating substance
Simple cuboidal epithelium	In ducts and secretory portions of small glands and in kidney tubules	Secretes and absorbs
Simple columnar epithelium	Ciliated tissues are in bronchi, uterine tubes, and uterus; smooth (nonciliated tissues) are in the digestive tract, bladder	Absorbs; it also secretes mucous and enzymes
Pseudostratified columnar epithelium	Ciliated tissue lines the trachea and much of the upper respiratory tract	Secretes mucus; ciliated tissue moves mucus
Stratified squamous epithelium	Lines the esophagus, mouth, and vagina	Protects against abrasion
Stratified cuboidal epithelium	Sweat glands, salivary glands, and the mammary glands	Protective tissue
Stratified columnar epithelium	The male urethra and the ducts of some glands	Secretes and protects
Transitional epithelium	Lines the bladder, urethra, and the ureters	Allows the urinary organs to expand and stretch

Summary showing different epithelial cells/tissues and their characteristics

The three principal shapes associated with epithelial cells are—squamous, cuboidal and columnar.

- Squamous epithelium has cells that are wider than their height (flat and scale-like).

- Cuboidal epithelium has cells whose height and width are approximately the same (cube shaped).

- Columnar epithelium has cells taller than they are wide (column-shaped).

By layer, epithelium is classed as either simple epithelium, only one cell thick (unilayered) or stratified epithelium as stratified squamous epithelium, stratified cuboidal epithelium, and stratified columnar epithelium that are two or more cells thick (multi-layered), and both types of layering can be made up of any of the cell shapes. However, when taller simple columnar epithelial cells are viewed in cross section showing several nuclei appearing at different heights, they can be confused with stratified epithelia. This kind of epithelium is therefore described as pseudostratified columnar epithelium.

Transitional epithelium has cells that can change from squamous to cuboidal, depending on the amount of tension on the epithelium.

Simple Epithelium

Simple epithelium is a single layer of cells with every cell in direct contact with the basement membrane that separates it from the underlying connective tissue. In general, it is found where absorption and filtration occur. The thinness of the epithelial barrier facilitates these processes.

In general, simple epithelial tissues are classified by the shape of their cells. The four major classes of simple epithelium are: (1) simple squamous; (2) simple cuboidal; (3) simple columnar; (4) pseudostratified.

(1) Simple squamous; which is found lining areas where passive diffusion of gases occur. e.g. skin, walls of capillaries, linings of the pericardial, pleural,and peritoneal cavities, as well as the linings of the alveoli of the lungs.

(2) Simple cuboidal: these cells may have secretory, absorptive, or excretory functions. examples include small collecting ducts of kidney, pancreas, and salivary gland.

(3) Simple columnar; cells can be secretory, absorptive, or excretory; Simple columnar epithelium can be ciliated or non-ciliated; ciliated columnar is found in the female reproductive tract and uterus. Non-ciliated epithelium can also possess microvilli.

(4) Pseudostratified columnar epithelium; can be ciliated or non-ciliated. The ciliated type is also called respiratory epithelium as it is almost exclusively confined to the larger respiratory airways of the nasal cavity, trachea and bronchi.

Stratified Epithelium

Stratified epithelium differs from simple epithelium in that it is multilayered. It is therefore found where body linings have to withstand mechanical or chemical insult such that layers can be abraded and lost without exposing subepithelial layers. Cells flatten as the layers become more apical, though in their most basal layers the cells can be squamous, cuboidal or columnar.

Stratified epithelia (of columnar, cuboidal or squamous type) can have the following specializations:

Specialization	Description
Keratinized	In this particular case, the most apical layers (exterior) of cells are dead and lose their nucleus and cytoplasm, instead contain a tough, resistant protein called keratin. This specialization makes the epithelium waterproof, so is found in the mammalian skin. The lining of the esophagus is an example of a non-keratinized or "moist" stratified epithelium.
Parakeratinized	In this case, the most apical layers of cells are filled with keratin, but they still retain their nuclei. These nuclei are pyknotic, meaning that they are highly condensed. Parakeratinized epithelium is sometimes found in the oral mucosa and in the upper regions of the esophagus.
Transitional	Transitional epithelia are found in tissues that stretch and it can appear to be stratified cuboidal when the tissue is not stretched or stratified squamous when the organ is distended and the tissue stretches. It is sometimes called urothelium since it is almost exclusively found in the bladder, ureters and urethra.

Cell Types

The basic cell types are squamous, cuboidal, and columnar classed by their shape.

Type	Description
Squamous	Squamous cells have the appearance of thin, flat plates that can look polygonal when viewed from above. Their name comes from *squāma*, Latin for scale – as on fish or snake skin. The cells fit closely together in tissues; providing a smooth, low-friction surface over which fluids can move easily. The shape of the nucleus usually corresponds to the cell form and helps to identify the type of epithelium. Squamous cells tend to have horizontally flattened, nearly oval shaped nuclei because of the thin flattened form of the cell. Squamous epitheliium is found lining surfaces such as the skin, and alveoli in the lung, enabling simple passive diffusion as also found in the alveolar epithelium in the lungs. Specialized squamous epithelium also forms the lining of cavities such as in blood vessels, as endothelium and in the pericardium, as mesothelium and in other body cavities.
Cuboidal	Cuboidal epithelial cells have a cube-like shape and appear square in cross-section. The cell nucleus is large, spherical and is in the centre of the cell. Cuboidal epithelium is commonly found in secretive tissue such as the exocrine glands, or in absorptive tissue such as the pancreas, the lining of the kidney tubules as well as in the ducts of the glands. The germinal epithelium that covers the female ovary, and the germinal epithelium that lines the walls of the seminferous tubules in the testes are also of the cuboidal type. Cuboidal cells provide protection and may be active in pumping material in or out of the lumen, or passive depending on their location and specialisation. Simple cuboidal epithelium commonly differentiates to form the secretory and duct portions of glands. Stratified cuboidal epithelium protects areas such as the ducts of sweat glands, mammary glands, and salivary glands.
Columnar	Columnar epithelial cells are elongated and column-shaped and have a height of at least four times their width. Their nuclei are elongated and are usually located near the base of the cells. Columnar epithelium forms the lining of the stomach and intestines. The cells here may possess microvilli for maximising the surface area for absorption and these microvilli may form a brush border. Other cells may be ciliated to move mucus in the function of mucociliary clearance. Other ciliated cells are found in the fallopian tubes, the uterus and central canal of the spinal cord. Some columnar cells are specialized for sensory reception such as in the nose, ears and the taste buds. Hair cells in the inner ears have stereocilia which are similar to microvilli. Goblet cells are modified columnar cells and are found between the columnar epithelial cells of the duodenum. They secrete mucus, which acts as a lubricant. Single-layered non-ciliated columnar epithelium tends to indicate an absorptive function. Stratified columnar epithelium is rare but is found in lobar ducts in the salivary glands, the eye, pharynx and sex organs. This consists of a layer of cells resting on at least one other layer of epithelial cells which can be squamous, cuboidal, or columnar.
Pseudostratified	These are simple columnar epithelial cells whose nuclei appear at different heights, giving the misleading (hence "pseudo") impression that the epithelium is stratified when the cells are viewed in cross section. Ciliated pseudostratified epithelial cells have cilia. Cilia are capable of energy dependent pulsatile beating in a certain direction through interaction of cytoskeletal microtubules and connecting structural proteins and enzymes. In the respiratory tract the wafting effect produced causes mucus secreted locally by the goblet cells (to lubricate and to trap pathogens and particles) to flow in that direction (typically out of the body). Ciliated epithelium is found in the airways (nose, bronchi), but is also found in the uterus and Fallopian tubes, where the cilia propel the ovum to the uterus.

Structure

Cells of epithelial tissue are tightly packed and form a continuous sheet. They have almost no intercellular spaces. All epithelia is usually separated from underlying tissues by an extracellular fibrous basement membrane. The lining of the mouth, lung alveoli and kidney tubules all are made of epithelial tissue. The lining of the blood and lymphatic vessels are of a specialised form of epithelium called endothelium.

Location

Epithelium lines both the outside (skin) and the inside cavities and lumina of bodies. The outer-most layer of human skin is composed of dead stratified squamous, keratinized epithelial cells.

Tissues that line the inside of the mouth, the esophagus and part of the rectum are composed of nonkeratinized stratified squamous epithelium. Other surfaces that separate body cavities from the outside environment are lined by simple squamous, columnar, or pseudostratified epithelial cells. Other epithelial cells line the insides of the lungs, the gastrointestinal tract, the reproductive and urinary tracts, and make up the exocrine and endocrine glands. The outer surface of the cornea is covered with fast-growing, easily regenerated epithelial cells. A specialised form of epithelium – endothelium forms the inner lining of blood vessels and the heart, and is known as vascular endothelium, and lining lymphatic vessels as lymphatic endothelium. Another type, mesothelium, forms the walls of the pericardium, pleurae, and peritoneum.

In arthropods, the integument, or external "skin", consists of a single layer of epithelial ectoderm from which arises the cuticle, an outer covering of chitin the rigidity of which varies as per its chemical composition.

Basement Membrane

Epithelial tissue rests on a basement membrane, which acts as a scaffolding on which epithelium can grow and regenerate after injuries. Epithelial tissue has a nerve supply, but no blood supply and must be nourished by substances diffusing from the blood vessels in the underlying tissue. The basement membrane acts as a selectively permeable membrane that determines which substances will be able to enter the epithelium.

Cell Junctions

Cell junctions are especially abundant in epithelial tissues. They consist of protein complexes and provide contact between neighbouring cells, between a cell and the extracellular matrix, or they build up the paracellular barrier of epithelia and control the paracellular transport.

Cell junctions are the contact points between plasma membrane and tissue cells. There are mainly 5 different types of cell junctions: tight junctions, adherens junctions, desmosomes, hemidesmosomes, and gap junctions. Tight junctions are a pair of trans-membrane protein fused on outer plasma membrane. Adherens junctions are a plaque (protein layer on the inside plasma membrane) which attaches both cells' microfilaments. Desmosomes attach to the microfilaments of cytoskeleton made up of keratin protein. Hemidesmosomes resemble desmosomes on a section. They are made up of the integrin (a transmembrane protein) instead of cadherin. They attach the epithelial cell to the basement membrane. Gap junctions connect the cytoplasm of two cells and are made up of proteins called connexins (six of which come together to make a connexon).

Development

Epithelial tissues are derived from all of the embryological germ layers:

- from ectoderm (e.g., the epidermis);

- from endoderm (e.g., the lining of the gastrointestinal tract);

- from mesoderm (e.g., the inner linings of body cavities).

However, it is important to note that pathologists do not consider endothelium and mesothelium (both derived from mesoderm) to be true epithelium. This is because such tissues present very different pathology. For that reason, pathologists label cancers in endothelium and mesothelium sarcomas, whereas true epithelial cancers are called carcinomas. Also, the filaments that support these mesoderm-derived tissues are very distinct. Outside of the field of pathology, it is, in general, accepted that the epithelium arises from all three germ layers.

Function

The primary functions of epithelial tissues are: (1) to protect the tissues that lie beneath it from radiation, desiccation, toxins, invasion by pathogens, and physical trauma; (2) the regulation and exchange of chemicals between the underlying tissues and a body cavity; (3) the secretion of hormones into the blood vascular system, and/or the secretion of sweat, mucus, enzymes, and other products that are delivered by ducts; (4) to provide sensation.

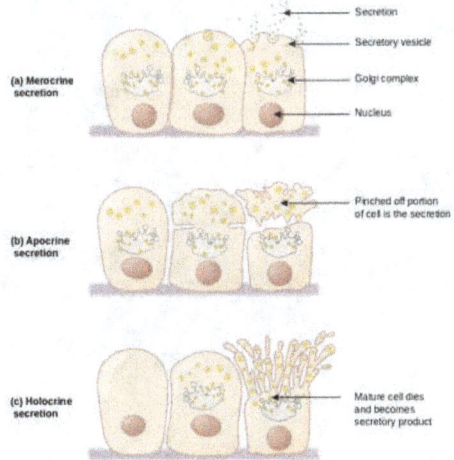

Forms of secretion in glandular tissue

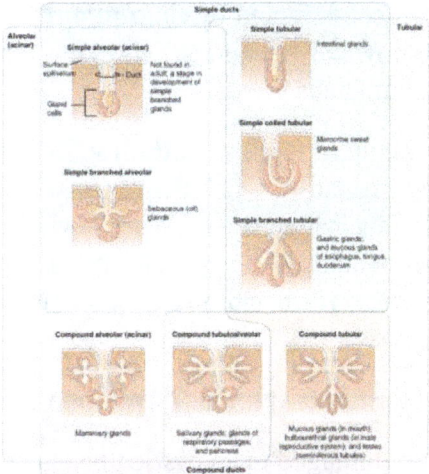

Different characteristics of glands of the body

Glandular Tissue

Glandular tissue is the type of epithelium that forms the glands from the infolding of epithelium and subsequent growth in the underlying connective tissue. There are two major classifications of glands: endocrine glands and exocrine glands. Endocrine glands secrete their product into the extracellular space where it is rapidly taken up by the blood vascular system. Exocrine glands secrete their products into a duct that then delivers the product to the lumen of an organ or onto the free surface of the epithelium.

Sensing the Extracellular Environment

"Some epithelial cells are ciliated, especially in respiratory epithelium, and they commonly exist as a sheet of polarised cells forming a tube or tubule with cilia projecting into the lumen." Primary cilia on epithelial cells provide chemosensation, thermoception, and mechanosensation of the extracellular environment by playing "a sensory role mediating specific signalling cues, including soluble factors in the external cell environment, a secretory role in which a soluble protein is released to have an effect downstream of the fluid flow, and mediation of fluid flow if the cilia are motile."

Clinical Significance

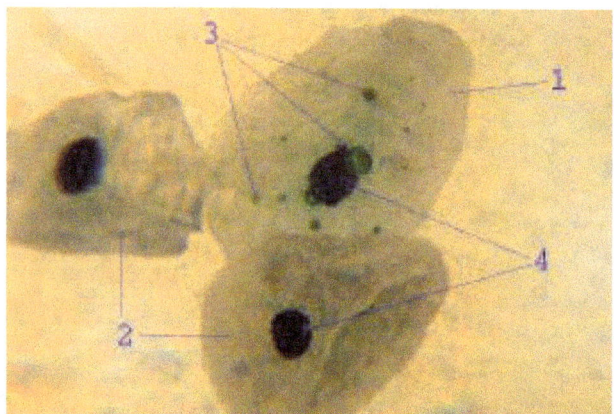

Epithelial cell infected with *Chlamydia pneumoniae*

The slide shows at (1) an epithelial cell infected by *Chlamydia pneumonia*; their inclusion bodies shown at (3); an uninfected cell shown at (2) and (4) showing the difference between an infected cell nucleus and an uninfected cell nucleus.

Epithelium grown in culture can be identified by examining its morphological characteristics. Epithelial cells tend to cluster together, and have a "characteristic tight pavementlike appearance". But this is not always the case, such as when the cells are derived from a tumor. In these cases, it is often necessary to use certain biochemical markers to make a positive identification. The intermediate filament proteins in the cytokeratin group are almost exclusively found in epithelial cells, and so are often used for this purpose.

Carcinomas develop in epithelial tissues.(Sarcomas develop in connective tissue).

When epithelial cells or tissues are damaged from cystic fibrosis, it also damages the sweat glands, causing a frosty coating of the skin.

Etymology

The word *epithelium* uses the Greek roots (*epi*), "on" or "upon", and (*thēlē*), "nipple". Epithelium is so called because the name was originally used to describe the translucent covering of small "nipples" of tissue on the lip. The word has both mass and count senses; the plural form is epithelia.

Muscle Tissue

Muscle tissue is a soft tissue that composes muscles in animal bodies, and gives rise to muscles' ability to contract. This is opposed to other components or tissues in muscle such as tendons or perimysium. It is formed during embryonic development through a process known as myogenesis.

Muscle tissue varies with function and location in the body. In mammals the three types are: skeletal or striated muscle; smooth or non-striated muscle; and cardiac muscle, which is sometimes known as semi-striated. Smooth and cardiac muscle contracts involuntarily, without conscious intervention. These muscle types may be activated both through interaction of the central nervous system as well as by receiving innervation from peripheral plexus or endocrine (hormonal) activation. Striated or skeletal muscle only contracts voluntarily, upon influence of the central nervous system. Reflexes are a form of non-conscious activation of skeletal muscles, but nonetheless arise through activation of the central nervous system, albeit not engaging cortical structures until after the contraction has occurred.

The different muscle types vary in their response to neurotransmitters and endocrine substances such as acetyl-choline, noradrenalin, adrenalin, nitric oxide and among others depending on muscle type and the exact location of the muscle.

Sub-categorization of muscle tissue is also possible, depending on among other things the content of myoglobin, mitochondria, myosin ATPase etc.

Structure

Muscle (myocytes) are elongated cells ranging from several millimetres to about 10 centimetres in length and from 10 to 100 micrometres in width. These cells are joined together in tissues that may be either striated or smooth, depending on the presence or absence, respectively, of organized, regularly repeated arrangements of myofibrillar contractile proteins called myofilaments. Striated muscle is further classified as either skeletal or cardiac muscle. Striated muscle is typically subject to conscious control, while smooth muscle is not. Thus, muscle tissue can be described as being one of three different types:

- Skeletal muscle, striated in structure and under voluntary control, is anchored by tendons (or by aponeuroses at a few places) to bone and is used to effect skeletal movement such as locomotion and to maintain posture. (Though postural control is generally maintained as an unconscious reflex—the muscles responsible also react to conscious control like non-postural muscles.) An average adult male is made up of 42% of skeletal muscle and an average adult female is made up of 36% (as a percentage of body mass). It also has striations unlike smooth muscle.

- Smooth muscle, neither striated in structure nor under voluntary control, is found within the walls of organs and structures such as the esophagus, stomach, intestines, bronchi, uterus, urethra, bladder, blood vessels, and the arrector pili in the skin (in which it controls erection of body hair).

In vertebrates, there is a third muscle tissue recognized:

- Cardiac muscle (myocardium), found only in the heart, is a striated muscle similar in structure to skeletal muscle but not subject to voluntary control.

Cardiac and skeletal muscles are "striated" in that they contain sarcomeres and are packed into highly regular arrangements of bundles; smooth muscle has neither. While skeletal muscles are arranged in regular, parallel bundles, cardiac muscle connects at branching, irregular angles (called intercalated discs). Striated muscle contracts and relaxes in short, intense bursts, whereas smooth muscle sustains longer or even near-permanent contractions.

Comparison of Types

	smooth muscle	cardiac muscle	skeletal muscle
Anatomy			
Neuromuscular junction	none	none	present
Fibers	fusiform, short (<0.4 mm)	branching	cylindrical, long (<15 cm)
Mitochondria	few	numerous	many to few (by type)
Nuclei	1	1	>1
Sarcomeres	none	present, max. length 2.6 µm	present, max. length 3.7 µm
Syncytium	none (independent cells)	none (but functional as such)	present
Sarcoplasmic reticulum	little elaborated	moderately elaborated	highly elaborated
ATPase	little	moderate	abundant
Physiology			
Self-regulation	spontaneous action (slow)	yes (rapid)	none (requires nerve stimulus)
Response to stimulus	unresponsive	"all-or-nothing"	"all-or-nothing"
Action potential	yes	yes	yes
Workspace	Force/length curve is variable	the increase in the force/length curve	at the peak of the force/length curve
Response to stimulus			

Skeletal Muscle

Skeletal muscle is further divided into several subtypes:

- Type I, slow oxidative, slow twitch, or "red" muscle is dense with capillaries and is rich in mitochondria and myoglobin, giving the muscle tissue its characteristic red color. It can carry more oxygen and sustain aerobic activity.

  o Type I muscle fiber are sometimes broken down into Type I and Type Ic categories, as a result of recent research.

- Type II, fast twitch muscle, has three major kinds that are, in order of increasing contractile speed:

  o Type IIa, which, like slow muscle, is aerobic, rich in mitochondria and capillaries and appears red when deoxygenated.

  o Type IIx (also known as type IId), which is less dense in mitochondria and myoglobin. This is the fastest muscle type in humans. It can contract more quickly and with a greater amount of force than oxidative muscle, but can sustain only short, anaerobic bursts of activity before muscle contraction becomes painful (often incorrectly attributed to a build-up of lactic acid).

  o Type IIb, which is anaerobic, glycolytic, "white" muscle that is even less dense in mitochondria and myoglobin. In small animals like rodents this is the major fast muscle type, explaining the pale color of their flesh.

Striated skeletal muscle cells in microscopic view. The myofibers are the straight vertical bands; the horizontal striations (lighter and darker bands) that are visible result from differences in composition and density along the fibrils within the cells. The cigar-like dark patches beside the myofibers are muscle-cell nuclei.

Smooth Muscle

Smooth muscle is an involuntary non-striated muscle. It is divided into two subgroups: the single-unit (unitary) and multiunit smooth muscle. Within single-unit cells, the whole bundle or sheet contracts as a syncytium (i.e. a multinucleate mass of cytoplasm that is not separated into cells). Multiunit smooth muscle tissues innervate individual cells; as such, they allow for fine control and gradual responses, much like motor unit recruitment in skeletal muscle.

Smooth muscle is found within the walls of blood vessels (such smooth muscle specifically being termed vascular smooth muscle) such as in the tunica media layer of large (aorta) and small arteries, arterioles and veins. Smooth muscle is also found in lymphatic vessels, the urinary bladder, uterus (termed uterine smooth muscle), male and female reproductive tracts, gastro-intestinal tract, respiratory tract, arrector pili of skin, the ciliary muscle, and iris of the eye. The structure and function is basically the same in smooth muscle cells in different organs, but the inducing stimuli differ substantially, in order to perform individual effects in the body at individual times. In addition, the glomeruli of the kidneys contain smooth muscle-like cells called mesangial cells.

Cardiac Muscle

Cardiac muscle is involuntary, striated muscle that is found in the walls and histological foundation of the heart, specifically the myocardium. Cardiac muscle is one of three major types of muscle, the others being skeletal and smooth muscle. These three types of muscle all form in the process of myogenesis. The cells that constitute cardiac muscle, called cardiomyocytes or myocardiocytes, predominantly contain only one nucleus, although populations with two to four nuclei do exist. The myocardium is the muscle tissue of the heart, and forms a thick middle layer between the outer epicardium layer and the inner endocardium layer.

Coordinated contractions of cardiac muscle cells in the heart propel blood out of the atria and ventricles to the blood vessels of the left/body/systemic and right/lungs/pulmonary circulatory systems. This complex mechanism illustrates systole of the heart.

Cardiac muscle cells, unlike most other tissues in the body, rely on an available blood and electrical supply to deliver oxygen and nutrients and remove waste products such as carbon dioxide. The coronary arteries help fulfill this function.

Function

Skeletal Muscle

1. They carry out movements of the body.

2. They support the body.

3. They maintain the posture of the body.

Smooth Muscle

It is responsible for the contractility of hollow organs, such as blood vessels, the gastrointestinal tract, the bladder.

Cardiac Muscle

Cardiac muscle is the muscle of the heart. It is self-contracting, autonomically regulated and must continue to contract in rhythmic fashion for the whole life of the organism. Hence it has special features.

References

- Sternini, C (February 2007). "Taste receptors in the gastrointestinal tract. IV. Functional implications of bitter taste receptors in gastrointestinal chemosensing.". American Journal of Physiology. Gastrointestinal and Liver Physiology. 292 (2): G457–61. PMID 17095755. doi:10.1152/ajpgi.00411.2006

- al.], consultants Daniel Albert ... [et (2012). Dorland's illustrated medical dictionary. (32nd ed.). Philadelphia, PA: Saunders/Elsevier. p. 321. ISBN 978-1-4160-6257-8

- Tamarkin, Dawn. "Myofibril Composition". www.stcc.edu/faculty/webpages.asp. STCC Foundation Press. Retrieved 12 February 2015

- Holstein T.; Tardent P. (1984). "An ultrahigh-speed analysis of exocytosis: nematocyst discharge". Science. 223 (4638): 830–833. PMID 6695186. doi:10.1126/science.6695186

- Krause, WJ; Yamada, J; Cutts, JH (June 1985). "Quantitative distribution of enteroendocrine cells in the gastrointestinal tract of the adult opossum, Didelphis virginiana." (PDF). Journal of Anatomy. 140 (4): 591–605. PMC 1165084. PMID 4077699

- MacIntosh, Brian R.; Gardiner, Phillip F.; McComas, Alan J. (2006). Skeletal Muscle: Form and Function. Human Kinetics. ISBN 978-0-7360-4517-9

- "M Cell Differentiation: Distinct Lineage or Phenotypic Transition? Salmonella Provides Answers". Cell Host & Microbe. 12: 607–609. doi:10.1016/j.chom.2012.11.003. Retrieved 2016-01-16

- Brinkman D, Burnell J (November 2007). "Identification, cloning and sequencing of two major venom proteins from the box jellyfish, Chironex fleckeri". Toxicon. 50 (6): 850–60. PMID 17688901. doi:10.1016/j.toxicon.2007.06.016

- Miller H.; Zhang J.; Kuolee R.; Patel G.B.; Chen W. (2007). "Intestinal M cells: the fallible sentinels?". World Journal of Gastroenterology. 13: 1477–1486. doi:10.3748/wjg.v13.i10.1477

- Hayes, A. Wallace (2007), Principles and Methods of Toxicology (5th, revised ed.), CRC Press, p. 1547, ISBN 9781420005424

- McCloud, Aaron (30 November 2011). "Build Fast Twitch Muscle Fibers". Complete Strength Training. Retrieved 30 November 2011

- Warner, RR (May 2005). "Enteroendocrine tumors other than carcinoid: a review of clinically significant advances.". Gastroenterology. 128 (6): 1668–84. PMID 15887158. doi:10.1053/j.gastro.2005.03.078

- Fakhrejahani, Elham; Toi, Masakazu (2012). "Tumor Angiogenesis: Pericytes and Maturation Are Not to Be Ignored". Journal of Oncology. 2012: 1–10. PMC 3191787. PMID 22007214. doi:10.1155/2012/261750

- Dore-Duffy, Paula; Cleary, Kristen (2011). "The Blood-Brain and Other Barriers". Methods in Molecular Biology. 686: 49–68. ISBN 978-1-60761-937-6. PMID 21082366. doi:10.1007/978-1-60761-938-3_2

- Nüchter Timm; Benoit Martin; Engel Ulrike; Özbek Suat; Holstein Thomas W (2006). "Nanosecond-scale kinetics of nematocyst discharge". Current Biology. 16 (9): R316–R318. PMID 16682335. doi:10.1016/j.cub.2006.03.089. Retrieved 2012-10-25

- Peppiatt, Claire M.; Howarth, Clare; Mobbs, Peter; Attwell, David (2006). "Bidirectional control of CNS capillary diameter by pericytes.". Nature. 443: 700–704. PMC 17618488.. PMID 17036005. doi:10.1038/nature05193

- Larsson, L; Edström, L; Lindegren, B; Gorza, L; Schiaffino, S (July 1991). "MHC composition and enzyme-histochemical and physiological properties of a novel fast-twitch motor unit type". The American Journal of Physiology. 261 (1 pt 1): C93–101. PMID 1858863. Retrieved 2006-06-11

- McConnell, Thomas H. (2006). The nature of disease: pathology for the health professions. Lippincott Williams & Wilkins. p. 55. ISBN 978-0-7817-5317-3

3

Sensory Organs and Nervous System in Animals

Lateral lines are found in aquatic animals. It helps them in sensing the movement and vibration which surrounds them in water. The other types of sensory organs and nervous systems explained are cercus, crista acustica, forked tongue, Mauthner cell, medullary command nucleus, etc. The topics elaborated in this chapter will help in gaining a better perspective about the subject matter.

Lateral Line

This is an example of lateral line of a fish.

The lateral line is a system of sense organs found in aquatic vertebrates, used to detect movement, vibration, and pressure gradients in the surrounding water. The sensory ability is achieved via modified epithelial cells, known as hair cells, which respond to displacement caused by motion and transduce these signals into electrical impulses via excitatory synapses. Lateral lines serve an important role in schooling behavior, predation, and orientation. For example, fish can use their lateral line system to follow the vortices produced by fleeing prey. Lateral lines are usually visible as faint lines of pores running lengthwise down each side, from the vicinity of the gill covers to the base of the tail. In some species, the receptive organs of the lateral line have been modified to function as electroreceptors, which are organs used to detect electrical impulses, and as such, these systems remain closely linked. Most amphibian larvae and some fully aquatic adult amphibians possess mechanosensitive systems comparable to the lateral line.

Function

The lateral line system allows the detection of movement, vibration, and pressure gradients in the water surrounding an animal, providing spatial awareness and the ability to navigate in the environment. This plays an essential role in orientation, predatory behavior, defense, and social schooling.

The small holes on the head of this Northern pike (*Esox lucius*) contain neuromasts of the lateral line system.

The lateral line system is necessary to detect vibrations made by prey, and to orient towards the source to begin predatory action. Fish are able to detect movement, produced either by prey or a vibrating metal sphere, and orient themselves toward the source before proceeding to make a predatory strike at it. This behavior persists even in blinded fish, but is greatly diminished when lateral line function was inhibited by $CoCl_2$ application. Cobalt chloride treatment results in the release of cobalt ions, disrupting ionic transport and preventing signal transduction in the lateral lines. These behaviors are dependent specifically on mechanoreceptors located within the canals of the lateral line.

The role mechanoreception plays in schooling behavior was demonstrated in a 1976 study. A school of *Pollachius virens* was established in a tank and individual fish were removed and subjected to different procedures before their ability to rejoin the school was observed. Fish that were experimentally blinded were able to reintegrate into the school, while fish with severed lateral lines were unable to reintegrate themselves. Therefore, reliance on functional mechanoreception, not vision, is essential for schooling behavior. A study in 2014 suggests that the lateral line system plays an important role in the behavior of Mexican blind cave fish (Astyanax mexicanus).

Anatomy

The major unit of functionality of the lateral line is the neuromast. The neuromast is a mechanoreceptive organ which allows the sensing of mechanical changes in water. There are two main varieties of neuromasts located in animals, canal neuromasts and superficial or freestanding neuromasts. Superficial neuromasts are located externally on the surface of the body, while canal neuromasts are located along the lateral lines in subdermal, fluid filled canals. Each neuromast consists of receptive hair cells whose tips are covered by a flexible and jellylike cupula. Hair cells typically possess both glutamatergic afferent connections and cholinergic efferent connections. The receptive hair cells are modified epithelial cells and typically possess bundles of 40-50 microvilli "hairs" which function as the mechanoreceptors. These bundles are organized in rough "staircases" of hairs of increasing length order. This use of mechanosensitive hairs is homologous to the functioning of hair cells in the auditory and vestibular systems, indicating a close link between these systems.

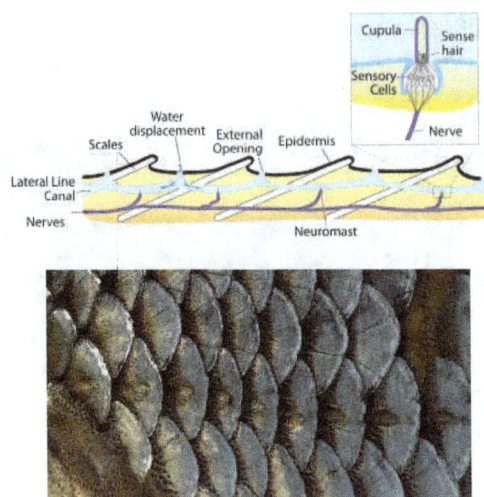

Some scales of the lateral line (center) of a *Rutilus rutilus*.

Hair cells utilize a system of transduction that uses rate coding in order to transmit the directionality of a stimulus. Hair cells of the lateral line system produce a constant, tonic rate of firing. As mechanical motion is transmitted through water to the neuromast, the cupula bends and is displaced. Varying in magnitude with the strength of the stimulus, shearing movement and deflection of the hairs is produced, either toward the longest hair or away from it. This results in a shift in the cell's ionic permeability, resulting from changes to open ion channels caused by the deflection of the hairs. Deflection towards the longest hair results in depolarization of the hair cell, increased neurotransmitter release at the excitatory afferent synapse, and a higher rate of signal transduction. Deflection towards the shorter hair has the opposite effect, hyperpolarizing the hair cell and producing a decreased rate of neurotransmitter release. These electrical impulses are then transmitted along afferent lateral neurons to the brain.

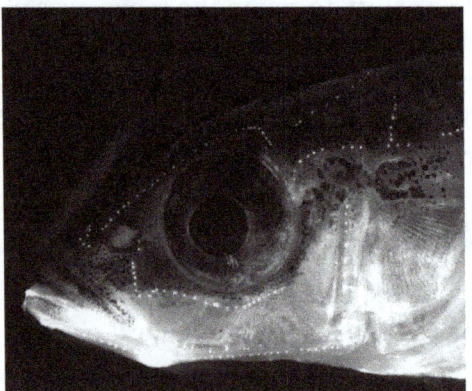

A *three-spined stickleback* with stained neuromasts

While both varieties of neuromasts utilize this method of transduction, the specialized organization of superficial and canal neuromasts allow them different mechanoreceptive capacities. Located at the surface of an animal's skin, superficial organs are exposed more directly to the external environment. Though these organs possess the standard "staircase" shaped hair bundles, overall the organization of the bundles within the organs is seemingly haphazard, incorporating various shapes and sizes of microvilli within bundles. This suggests a wide range of detection, potentially indicating a function of broad detection to determine the presence and magnitude of deflection caused by motion in the surrounding water. In contrast, the structure of canal organs allow canal neuromasts to be organized into a network system that allows more sophisticated mechanoreception, such as the detection of pressure differentials. As current moves across the pores of a canal, a pressure differential is created over the pores. As pressure on one pore exceeds that of another pore, the differential pushes down on the canal and causes flow in the canal fluid. This moves the cupula of the hair cells in the canal, resulting in a directional deflection of the hairs corresponding to the direction of the flow. This method allows the translation of pressure information into directional deflections which can be received and transduced by hair cells.

Electrophysiology

The mechanoreceptive hair cells of the lateral line structure are integrated into more complex circuits through their afferent and efferent connections. The synapses that directly participate in the transduction of mechanical information are excitatory afferent connections that utilize glutamate. However, a variety of different neuromast and afferent connections are possible, resulting in variation in mechanoreceptive properties. For instance, a series of experiments on the superficial neuromasts of *Porichthys notatus* revealed that neuromasts can exhibit a receptive specificity for particular frequencies of stimulation. Using an immobilized fish to prevent extraneous stimulation, a metal ball was vibrated at different frequencies. Utilizing single cell measurements with a microelectrode, responses were recorded and used to construct tuning curves, which revealed frequency preferences and two main afferent nerve types. One variety is attuned to collect mechanoreceptive information about acceleration, responding to stimulation frequencies between 30–200 Hz. The other type is sensitive to velocity information and is most receptive to stimulation below <30 Hz. This suggests a more intricate model of reception than was previously considered.

The efferent synapses to hair cells are inhibitory and utilize acetylcholine as a transmitter. They are crucial participants in a corollary discharge system designed to limit self-generated interference. When a fish moves, it creates disturbances in the water that could be detected by the lateral line system, potentially interfering with the detection of other biologically relevant signals. To prevent this, an afferent signal is sent to the hair cell upon motor action, resulting in inhibition which counteracts the excitation resulting from reception of the self-generated stimulation. This allows the fish to retain perception of motion stimuli without interference created by its own movements.

After signals transduced from the hair cells are transmitted along lateral neurons, they eventually reach the brain. Visualization methods have revealed that the area where these signals most often terminate is the medial octavolateralis nucleus (MON). It is likely that the MON plays an important role in the processing and integration of mechanoreceptive information. This has been supported through other experiments, such as the use of Golgi staining and microscopy by New & Coombs to demonstrate the presence of distinct cell layers within the MON. Distinct layers of basilar and non-basilar crest cells were identified within the deep MON. Drawing a comparison to similar cells in the closely related electrosensory lateral line lobe of electric fish, it seems to suggest possible computational pathways of the MON. The MON is likely involved in the integration of sophisticated excitatory and inhibitory parallel circuits in order to interpret mechanoreceptive information.

Sensory Systems in Fish

Most fish possess highly developed sense organs. Nearly all daylight fish have color vision that is at least as good as a human's. Many fish also have chemoreceptors that are responsible for extraordinary senses of taste and smell. Although they have ears, many fish may not hear very well. Most fish have sensitive receptors that form the lateral line system, which detects gentle currents and vibrations, and senses the motion of nearby fish and prey. Sharks can sense frequencies in the range of 25 to 50 Hz through their lateral line.

Fish orient themselves using landmarks and may use mental maps based on multiple landmarks or symbols. Fish behavior in mazes reveals that they possess spatial memory and visual discrimination.

Vision

Fish have a refractive index gradient within the lens of their eyes which compensates for spherical aberration. Unlike humans, most fish adjust focus by moving the lens closer or further from the retina.

Vision is an important sensory system for most species of fish. Fish eyes are similar to those of terrestrial vertebrates like birds and mammals, but have a more spherical lens. Their retinas generally have both rod cells and cone cells (for scotopic and photopic vision), and most species have colour vision. Some fish can see ultraviolet and some can see polarized light. Amongst jawless fish, the lamprey has well-developed eyes, while the hagfish has only primitive eyespots. Fish vision shows adaptation to their visual environment, for example deep sea fishes have eyes suited to the dark environment.

Fish and other aquatic animals live in a different light environment than terrestrial species. Water absorbs light so that with increasing depth the amount of light available decreases quickly. The optic properties of water also lead to different wavelengths of light being absorbed to different degrees, for example light of long wavelengths (e.g. red, orange) is absorbed quite quickly compared to light of short wavelengths (blue, violet), though ultraviolet light (even shorter wavelength than blue) is absorbed quite quickly as well. Besides these universal qualities of water, different bodies of water may absorb light of different wavelengths because of salts and other chemicals in the water.

Hearing

Hearing is an important sensory system for most species of fish. Hearing threshold and the ability to localize sound sources are reduced underwater, in which the speed of sound is faster than in air. Underwater hearing is by bone conduction, and localization of sound appears to depend on differences in amplitude detected by bone conduction. Aquatic animals such as fish, however, have a more specialized hearing apparatus that is effective underwater.

Fish can sense sound through their lateral lines and their otoliths (ears). Some fishes, such as some species of carp and herring, hear through their swim bladders, which function rather like a hearing aid.

Hearing is well-developed in carp, which have the Weberian organ, three specialized vertebral processes that transfer vibrations in the swim bladder to the inner ear.

Although it is hard to test sharks' hearing, they may have a sharp sense of hearing and can possibly hear prey many miles away. A small opening on each side of their heads (not the spiracle) leads directly into the inner ear through a thin channel. The lateral line shows a similar arrangement, and is open to the environment via a series of openings called lateral line pores. This is a reminder of the common origin of these two vibration- and sound-detecting organs that are grouped together as the acoustico-lateralis system. In bony fish and tetrapods the external opening into the inner ear has been lost.

Current Detection

Hair cells in fish are used to detect water movements around their bodies. These hair cells are embedded in a jelly-like protrusion called cupula. The hair cells therefore can not be seen and do not appear on the surface of skin.

The lateral line in fish and aquatic forms of amphibians is a detection system of water currents, consisting mostly of vortices. The lateral line is also sensitive to low-frequency vibrations. The mechanoreceptors are hair cells, the same mechanoreceptors for vestibular sense and hearing. It is used primarily for navigation, hunting, and schooling. The receptors of the electrical sense are modified hair cells of the lateral line system.

Fish and some aquatic amphibians detect hydrodynamic stimuli via a lateral line. This system consists of an array of sensors called neuromasts along the length of the fish's body. Neuromasts can be free-standing (superficial neuromasts) or within fluid-filled canals (canal neuromasts). The sensory cells within neuromasts are polarized hair cells contained within a gelatinous cupula. The cupula, and the stereocilia within, are moved by a certain amount depending on the movement of the surrounding water. Afferent nerve fibers are excited or inhibited depending on whether the hair cells they arise from are deflected in the preferred or opposite direction. Lateral line receptors form somatotopic maps within the brain informing the fish of amplitude and direction of flow at different points along the body. These maps are located in the medial octavolateral nucleus (MON) of the medulla and in higher areas such as the torus semicircularis.

Pressure Detection

Pressure detection uses the organ of Weber, a system consisting of three appendages of vertebrae transferring changes in shape of the gas bladder to the middle ear. It can be used to regulate the buoyancy of the fish. Fish like the weather fish and other loaches are also known to respond to low pressure areas but they lack a swim bladder.

Chemoreception

The shape of the hammerhead shark's head may enhance olfaction by spacing the nostrils further apart.

Sharks have keen olfactory senses, located in the short duct (which is not fused, unlike bony fish) between the anterior and posterior nasal openings, with some species able to detect as little as one part per million of blood in seawater.

Sharks have the ability to determine the direction of a given scent based on the timing of scent detection in each nostril. This is similar to the method mammals use to determine direction of sound.

They are more attracted to the chemicals found in the intestines of many species, and as a result often linger near or in sewage outfalls. Some species, such as nurse sharks, have external barbels that greatly increase their ability to sense prey.

The MHC genes are a group of genes present in many animals and important for the immune system; in general, offspring from parents with differing MHC genes have a stronger immune system. Fish are able to smell some aspect of the MHC genes of potential sex partners and prefer partners with MHC genes different from their own.

Salmon have a strong sense of smell. Speculation about whether odours provide homing cues, go back to the 19th century. In 1951, Hasler hypothesised that, once in vicinity of the estuary or entrance to its birth river, salmon may use chemical cues which they can smell, and which are unique to their natal stream, as a mechanism to home onto the entrance of the stream. In 1978, Hasler and his students convincingly showed that the way salmon locate their home rivers with such precision was indeed because they could recognise its characteristic smell. They further demonstrated that the smell of their river becomes imprinted in salmon when they transform into smolts, just before they migrate out to sea. Homecoming salmon can also recognise characteristic smells in tributary streams as they move up the main river. They may also be sensitive to characteristic pheromones given off by juvenile conspecifics. There is evidence that they can "discriminate between two populations of their own species".

Electroreception and Magnetoreception

Electroreception, or electroception, is the ability to detect electric fields or currents. Some fish, such as catfish and sharks, have organs that detect weak electric potentials on the order of millivolts. Other fish, like the South American electric fishes Gymnotiformes, can produce weak electric currents, which they use in navigation and social communication. In sharks, the ampullae of

Lorenzini are electroreceptor organs. They number in the hundreds to thousands. Sharks use the ampullae of Lorenzini to detect the electromagnetic fields that all living things produce. This helps sharks (particularly the hammerhead shark) find prey. The shark has the greatest electrical sensitivity of any animal. Sharks find prey hidden in sand by detecting the electric fields they produce. Ocean currents moving in the magnetic field of the Earth also generate electric fields that sharks can use for orientation and possibly navigation.

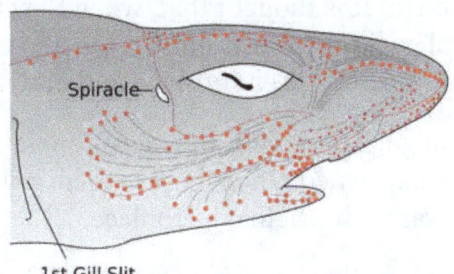

Electromagnetic field receptors (ampullae of Lorenzini) and motion detecting canals in the head of a shark

Electric field proximity sensing is used by the electric catfish to navigate through muddy waters. These fish make use of spectral changes and amplitude modulation to determine factors such shape, size, distance, velocity, and conductivity. The abilities of the electric fish to communicate and identify sex, age, and hierarchy within the species are also made possible through electric fields. EF gradients as low as 5nV/cm can be found in some saltwater weakly electric fish.

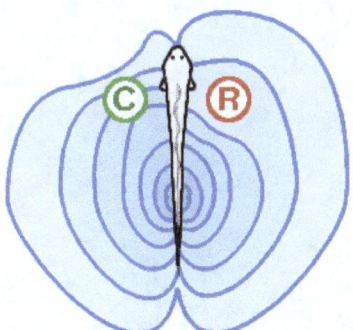

Active electrolocation. Conductive objects concentrate the field and resistive objects spread the field.

The paddlefish (*Polyodon spathula*) hunts plankton using thousands of tiny passive electroreceptors located on its extended snout, or rostrum. The paddlefish is able to detect electric fields that oscillate at 0.5–20 Hz, and large groups of plankton generate this type of signal.

Electric fishes use an active sensory system to probe the environment and create active electrodynamic imaging.

In 1973, it was shown that Atlantic salmon have conditioned cardiac responses to electric fields with strengths similar to those found in oceans. "This sensitivity might allow a migrating fish to align itself upstream or downstream in an ocean current in the absence of fixed references."

Magnetoception, or magnetoreception, is the ability to detect the direction one is facing based on the Earth's magnetic field. In 1988, researchers found iron, in the form of single domain magnetite, resides in the skulls of sockeye salmon. The quantities present are sufficient for magnetoception.

Fish Navigation

Salmon regularly migrate thousands of miles to and from their breeding grounds.

Salmon spend their early life in rivers, and then swim out to sea where they live their adult lives and gain most of their body mass. When they have matured, they return to the rivers to spawn. Usually they return with uncanny precision to the natal river where they were born, and even to the very spawning ground of their birth. It is thought that, when they are in the ocean, they use magnetoception to locate the general position of their natal river, and once close to the river, that they use their sense of smell to home in on the river entrance and even their natal spawning ground.

After several years wandering huge distances in the ocean, most surviving salmon return to the same natal rivers where they were spawned. Then most of them swim up the rivers until they reach the very spawning ground that was their original birthplace.

There are various theories about how this happens. One theory is that there are geomagnetic and chemical cues which the salmon use to guide them back to their birthplace. The fish may be sensitive to the Earth's magnetic field, which could allow the fish to orient itself in the ocean, so it can navigate back to the estuary of its natal stream.

Pain

Hooked sailfish

Experiments done by William Tavolga provide evidence that fish have pain and fear responses. For instance, in Tavolga's experiments, toadfish grunted when electrically shocked and over time they came to grunt at the mere sight of an electrode.

In 2003, Scottish scientists at the University of Edinburgh and the Roslin Institute concluded that rainbow trout exhibit behaviors often associated with pain in other animals. Bee venom and acetic acid injected into the lips resulted in fish rocking their bodies and rubbing their lips along the sides and floors of their tanks, which the researchers concluded were attempts to relieve pain, similar to what mammals would do. Neurons fired in a pattern resembling human neuronal patterns.

Professor James D. Rose of the University of Wyoming claimed the study was flawed since it did not provide proof that fish possess "conscious awareness, particularly a kind of awareness that is meaningfully like ours". Rose argues that since fish brains are so different from human brains, fish are probably not conscious in the manner humans are, so that reactions similar to human reactions to pain instead have other causes. Rose had published a study a year earlier arguing that fish cannot feel pain because their brains lack a neocortex. However, animal behaviorist Temple Grandin argues that fish could still have consciousness without a neocortex because "different species can use different brain structures and systems to handle the same functions."

Animal welfare advocates raise concerns about the possible suffering of fish caused by angling. Some countries, such as Germany have banned specific types of fishing, and the British RSPCA now prosecutes individuals who are cruel to fish.

Sensory Organs of Gastropods

The sensory organs of gastropods (snails and slugs) include olfactory organs, eyes, statocysts and mechanoreceptors. Gastropods have no sense of hearing.

Olfactory Organs

Rhinophores of *Aplysia californica*.

In terrestrial gastropods the most important sensory organs are the olfactory organs which are located on the tips of the 4 tentacles.

In opisthobranch marine gastropods, the chemosensory organs are two protruding structures on top of the head. These are known as rhinophores.

The upper pair of tentacles on the head of the edible snail *Helix pomatia* have eyes, but the main sensory organs are sensory neurons for olfaction in the epithelium of the tentacles.

An opisthobranch sea slug *Navanax inermis* has chemoreceptors on the sides of its mouth to track mucopolysaccharides in the slime trails of prey, and of potential mates.

The freshwater snail *Bithynia tentaculata* is capable of detecting the presence of molluscivorous (mollusk-eating) leeches through chemoreception, and of closing its operculum to avoid predation.

The deepwater snail *Bathynerita naticoidea* can detect mussel beds containing the mussel *Bathymodiolus childressi*, because it is attracted to water that has cues in it from this species of mussel.

Some terrestrial gastropods can track the odor of food using their tentacles (tropotaxis) and the wind (anemotaxis).

Eyes

In terrestrial pulmonate gastropods, eye spots are present at the tips of the tentacles in the Stylommatophora or at the base of the tentacles in the Basommatophora. These eye spots range from simple ocelli that cannot project an image (simply distinguishing light and dark), to more complex pit and even lens eyes. Vision is not the most important requirement in terrestrial gastropods, because they are mainly nocturnal animals.

Some gastropods, for example the freshwater Apple snails (family Ampullariidae) and marine species of genus *Strombus* can completely regenerate their eyes. The gastropods in both of these families have lens eyes.

Morphological sequence of different types of multicellular eyes exemplified by gastropod eyes:

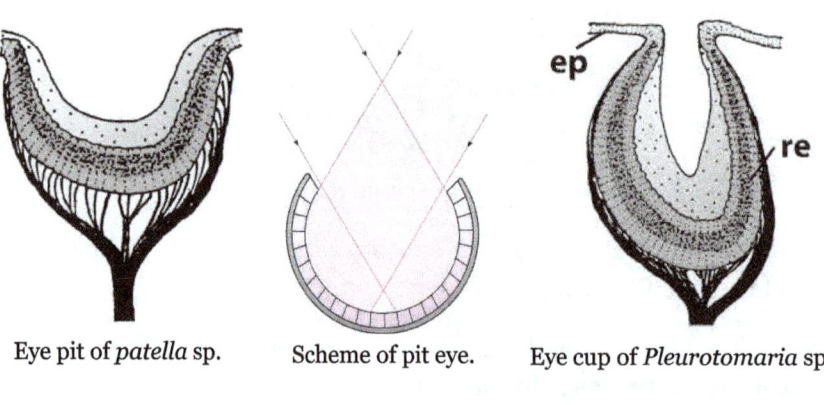

Eye pit of *patella* sp.   Scheme of pit eye.   Eye cup of *Pleurotomaria* sp.

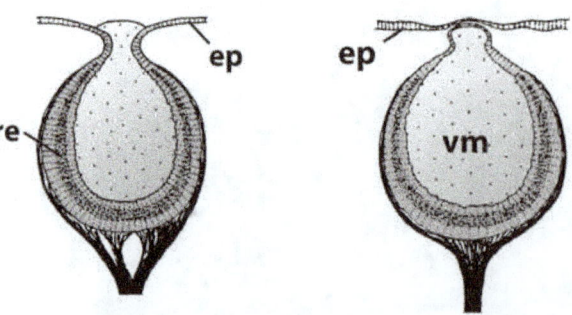

Pinhole eye of *Haliotis* sp.   Closed eye of *Turbo coronatus*.

Lens Eyes

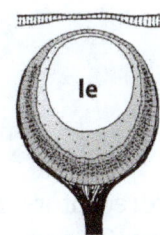

Lens eye of *Bolinus brandaris*.

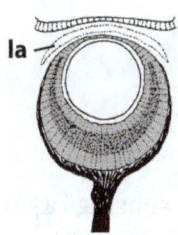

Lens eye of *Nucella lapillus*.

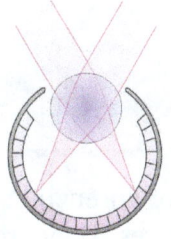

Scheme of lens eye.

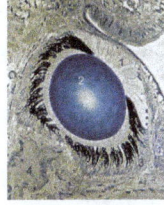

Eye of a snail.

1 - anterior chamber,

2 - lens,

3 - retina,

4 - optic nerve.

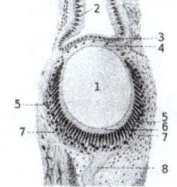

Drawing of cross section of the eye of *Helix pomatia*.

1 - lens

2 - olfactory epithelium

3 - corneal epithelium

4 - corneal endothelium

5 - retina

6 - layer with rod cells

7 - fibrous connective tissue layer

8 - nerve of the eye

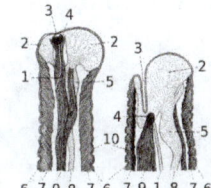

Drawing of cross sections of the extracted tentacle (left) and constricted tentacle (right) with and eye of *Helix pomatia*.

1 - nerve of an eye

4 - eye

5 - tentacle ganglion

6 - epidermis

8 - nerve of an tentacle

9 - retractor muscle

Well-developed lens eye of *Eustrombus gigas* on eyestalk has a black iris. There is a small tentacle on the eyestalk also.

Statocysts

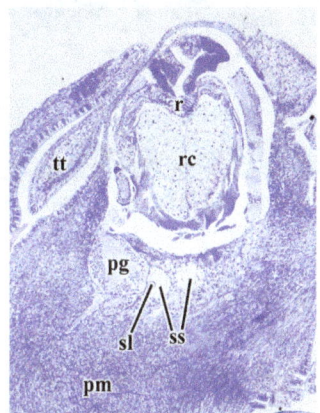

Statocysts (ss) and statolith (sl) inside the head of *Gigantopelta chessoia*.

In the statocysts of *Haliotis asinina* was found the expression of a conserved gene (Pax-258 gene), which is also important for forming structures for balance in eumetazoans.

Ampullae of Lorenzini

The ampullae of Lorenzini are special sensing organs called electroreceptors, forming a network of jelly-filled pores. They are mostly discussed as being found in cartilaginous fish (sharks, rays, and chimaeras); however, they are also reported to be found in Chondrostei such as reedfish and sturgeon. Lungfish have also been reported to have them. Teleosts have re-evolved a different type of electroreceptors. They were first described by Stefano Lorenzini in 1678.

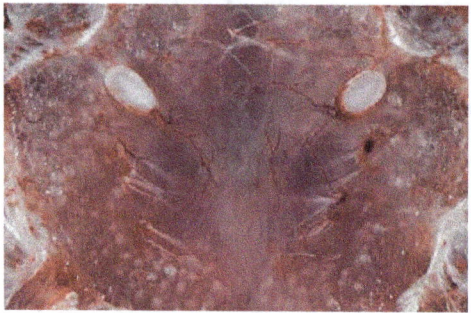
Inner view of Ampullae of Lorenzini

These sensory organs help fish to sense electric fields in the water. Each ampulla consists of a jelly-filled canal opening to the surface by a pore in the skin and ending blindly in a cluster of small pockets full of special jelly. The ampullae are mostly clustered into groups inside the body, each cluster having ampullae connecting with different parts of the skin, but preserving a left-right symmetry. The canal lengths vary from animal to animal, but the distribution of the pores is generally specific to each species. The ampullae pores are plainly visible as dark spots in the skin. They provide fish with an additional sense capable of detecting electric and magnetic fields as well as temperature gradients.

Pores with ampullae of Lorenzini in snout of Tiger shark

Electric Field Sensing Ability

The ampullae detect electric fields in the water, or more precisely the potential difference between the voltage at the skin pore and the voltage at the base of the electroreceptor cells. A positive pore stimulus would decrease the rate of nerve activity coming from the electroreceptor cells, and a negative pore stimulus would increase the rate of nerve activity coming from the electroreceptor cells. Each ampulla contains a single layer of cells that contains electrically excitable receptor cells separated by supporting cells. The cells are connected by apical tight junctions so that no current leaks between the cells. The apical faces of the receptor cells have a small surface area with a high concentration of voltage dependent calcium channels and calcium activated potassium channels. Because the canal wall has a very high resistance, all of the voltage difference between the pore of the canal and the ampulla is dropped across the receptor epithelium which is about 50 microns thick. Because the basal membranes of the receptor cells have a lower resistance, most of the voltage is dropped across the apical faces which are excitable and are poised at threshold. Inward calcium current across the receptor cells depolarizes the basal faces causing presynaptic calcium release and release of excitatory transmitter onto the afferent nerve fibers. One of the first descriptions of calcium activated potassium channels was based on studies of the ampulla of Lorenzini in the skate. Large conductance calcium activated potassium channels (BK channels) have recently been demonstrated in the ampulla by cloning.

Sharks may be more sensitive to electric fields than any other animal, with a threshold of sensitivity as low as 5 nV/cm. That is 5/1,000,000,000 of a volt measured in a centimeter-long ampulla. All living creatures produce an electrical field by muscle contractions, and a shark may pick up weak electrical stimuli from the muscle contractions of animals, particularly prey. On the other hand, the electrochemical fields generated by paralyzed prey were sufficient to elicit a feeding attack from sharks and rays in experimental tanks; therefore muscle contractions are not necessary to attract the animals. Sharks and rays can locate prey buried in the sand, or DC electric dipoles that simulate the main feature of the electric field of a prey buried in the sand.

Any moving conductor, such as sea water, induces an electric field when a magnetic field, such as the earth's is present. The electric fields induced in oceanic currents by the Earth's magnetic field are of the same order of magnitude as the electric fields that sharks and rays are capable of sensing. This could mean that sharks and rays can orient to the electric fields of oceanic currents, and use other sources of electric fields in the ocean for local orientation. Additionally, the electric field they induce in their bodies when swimming in the magnetic field of the Earth may enable them to sense their magnetic heading.

Behavioral studies have also provided evidence that sharks can detect changes in the geomagnetic field. In one experiment, sandbar sharks and scalloped hammerhead sharks were conditioned to associate a food reward with an artificial magnetic field. When the food reward was removed, the sharks continued to show a marked difference in behavior when the magnetic field was turned on as compared to when it was off.

Temperature Sensing Ability

Early in the 20th century, the purpose of the ampullae was not clearly understood, and electrophysiological experiments suggested a sensibility to temperature, mechanical pressure and possibly salinity. It was not until 1960 that the ampullae were clearly identified as specialized receptor organs for sensing electric fields. The ampullae may also allow the shark to detect changes in water temperature. Each ampulla is a bundle of sensory cells containing multiple nerve fibres. These fibres are enclosed in a gel-filled tubule which has a direct opening to the surface through a pore. The gel is a glycoprotein based substance with the same resistivity as seawater, and it has electrical properties similar to a semiconductor. This has been suggested as a mechanism by which temperature changes are transduced into an electrical signal that the shark may use to detect temperature gradients, although it is a subject of debate in scientific literature.

Material Properties

The hydrogel, which contains keratan sulfate in 97% water, has a conductivity of about 1.8 mS/cm, the highest known amongst biological materials.

Cercus

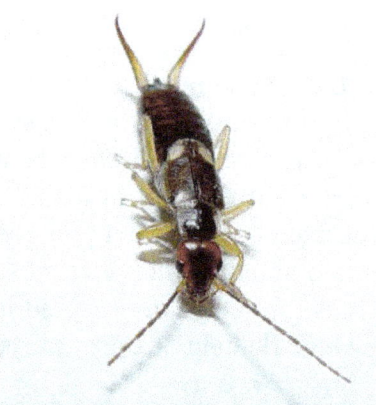

Earwig with large cerci (top)

Cerci (singular cercus) are paired appendages on the rear-most segments of many arthropods, including insects and symphylans. Many forms of cerci serve as sensory organs, but some serve as pinching weapons or as organs of copulation. In many insects, they simply may be functionless vestigial structures.

In basal arthropods, such as silverfish, the cerci originate from the eleventh abdominal segment. As segment eleven is reduced or absent in the majority of arthropods, in such cases, the cerci emerge from the tenth abdominal segment. It is not clear that other structures so named are homologous. In the Symphyla they are associated with spinnerets.

Morphology and Functions

Most cerci are segmented and jointed, or filiform (threadlike), but some take very different forms. Some Diplura, in particular *Japyx* species, have large, stout forcipate (pincer-like) cerci that they use in capturing their prey.

The Dermaptera, or earwigs, are well known for the forcipate cerci that most of them bear, though species in the suborders Arixeniina and Hemimerina do not. It is not clear how many of the Dermaptera use their cerci for anything but defense, but some definitely feed on prey caught with the cerci, much as the Japygidae do.

Crickets have particularly long cerci while other insects have cerci that are too small to be noticeable. However, it is not always obvious that small cerci are without function; they are rich in sensory cells and may be of importance in guiding copulation and oviposition.

Some insects such as mayflies, silverfish, and bristletails have an accompanying third central tail filament which extends from the tip of the abdomen. This is referred to as the *terminal filament* and is not regarded as a cercus.

Aphids have tube-like cornicles or siphunculi that are sometimes mistaken for cerci but are not morphologically related to cerci.

Evolutionary Origin

Like many insect body parts, including mandibles, antennae and styli, cerci are thought to have evolved from what were legs on the primal insect form; a creature that may have resembled a velvet worm, Symphylan or a centipede, worm-like with one pair of limbs for each segment behind the head or anterior tagma.

Crista Acustica

Crista acustica (also Siebold's organ, or crista acoustica) is a part of the hearing organ (tibial organ) in some insects (e.g. grasshoppers, crickets, katydids). It is a collection of sensory cells that form a crest (hence the name) on top of the hollow tube (the foreleg trachea) behind the hearing membrane (tympanum) on the legs of the insect.

The crista acustica is a transition from the intermedia organ (from the midline to the periphery), together which compose the tibial hearing organ (as opposed to the tympanal hearing organ).

The crista acustica is one of three fiddle-string-like organs or chordotonal organ in insects: the others are the intermediate organ and the supratympanal organ/subgenual organ. These chordotonal organs are actually collections of sensory cells sensitive to vibration (these cells are called scolopidia cells). Their cells are attached to the tube in the legs of the insects (the trachae, *"trah-key-ah"*). So when the tube vibrates, the cells vibrate. In the crista acustica, it turns out that different scolopidia cells are sensitive to different vibrations depending on the frequency of the vibration. Since these organs are sensitive to vibrations (due to changes in pressure. It turns out the cells closest to midline are the largest and sensitive to the lowest frequency (low frequencies having the largest wavelength), and the cells further out (distal) are smaller and sensitive to higher frequencies (since high frequencies have shorter wavelengths). This orderly arrangement of sensory cells gives the insect the ability to discriminate frequencies, much like the inner ear of mammals.

Forked Tongue

A forked tongue is a tongue split into two distinct tines at the tip; this is a feature common to many species of reptiles. Reptiles smell using the tip of their tongue, and a forked tongue allows them to sense from which direction a smell is coming. Sensing from both sides of the head and following trails based on chemical cues is called tropotaxis. It is unclear whether forked-tongued reptiles can actually follow trails or if this is just a hypothesis.

Forked tongue of a Carpet Python (*Morelia spilota mcdowelli*)

Forked tongues have evolved in these Squamate reptiles (lizards and snakes) for various purposes. The advantage to having a forked tongue is that more surface area is available for the chemicals to contact and the potential for tropotaxis. The tongue is flicked out of the mouth regularly to sample the chemical environment. This form of chemical sampling allows these animals to sense non-volatile chemicals, which cannot be detected by simply using the olfactory system. This increased ability to sense chemicals has allowed for heightened abilities to identify prey, recognize kin, choose mates, locate shelters, follow trails, and more.

Forked tongues have evolved multiple times in Squamates. It is unclear, based on the morphological and genetic evidence, where the exact points of change are from a notched tongue to a forked tongue, but it is believed that the change has happened two to four times. A common behavioral characteristic that has evolved in those with forked tongues is that they tend to be wide foragers.

Hummingbirds also have tongues that split at the tip. Galagos (bushbabies) have a secondary tongue, or sublingua, used for grooming, hidden under their first.

Usage as First Nations Cultural Term

The phrase "speaks with a forked tongue" means to deliberately say one thing and mean another or, to be hypocritical, or act in a duplicitous manner. In the longstanding tradition of many Native American tribes, "speaking with a forked tongue" has meant lying, and a person was no longer considered worthy of trust, once he had been shown to "speak with a forked tongue". This phrase was also adopted by Americans around the time of the Revolution, and may be found in abundant references from the early 19th century — often reporting on American officers who sought to convince the tribal leaders with whom they negotiated that they "spoke with a straight and not with a forked tongue" (as for example, President Andrew Jackson told the Creek Nation in 1829). According to one 1859 account, the native proverb that the "white man spoke with a forked tongue" originated as a result of the French tactic of the 1690s, in their war with the Iroquois, of inviting their enemies to attend a Peace Conference, only to be slaughtered or captured.

Literary Usage

There are appearances of the phrase "forked tongue" in English literature, either in reference to actual snakes' tongues, or as a metaphor for untruthfulness, such as a sermon by Lancelot Andrewes, who died in 1626:

"And he hath the art of cleaving. He shewed it in the beginning, when he made the Serpent, lingnam bisulcam, a forked tongue, to speake that, which was contrary to his knowledge and meaning, They should not die; and as hee did the Serpents, so hee can doe others."

The phrase also appears in Milton's Paradise Lost:

"According to his Doom: He would have spoke, But Hiss for Hiss return'd with forked Tongue To forked Tongue, for now were all transform'd"

Johnston's Organ

Johnston's organ is a collection of sensory cells found in the pedicel (the second segment) of the antennae in the Class Insecta. Johnston's organ detects motion in the flagellum (third and typically final antennal segment). It consists of scolopidia arrayed in a bowl shape, each of which contains a mechanosensory chordotonal neuron. The number of scolopidia varies between species. In homopterans, the Johnston's organs contain 25 - 79 scolopidia. The presence of Johnston's organ is a defining characteristic which separates the Class Insecta from the other hexapods belonging to the group Entognatha. Johnston's organ was named after the physician Christopher Johnston, father of the physician and Assyriologist Christopher Johnston.

Uses of the Johnston's Organ

In Fruit Flies

In the fruit fly *Drosophila melanogaster*, the Johnston's organ contains almost 480 sensory neurons. Distinct populations of neurons are activated differently by deflections of antennae caused by gravity or by vibrations caused by sound or air movement. This differential response allows the fly to distinguish between gravitational, mechanical, and acoustic stimuli.

The Johnston's organ of fruit flies can be used to detect air vibrations caused by the wingbeat frequency or courtship song of a mate. One function of the Johnston's organ is for detecting the wing beat frequency of a mate. Production of sound in air results in two energy components: the pressure component, which is changes in air pressure emanating away from the source of the sound; and the particle displacement component, which is the back and forth vibration of air particles oscillating in the direction of sound propagation. Particle displacement has greater energy loss than the pressure component, so the displacement component, called "near-field sound," is detectable only within one wavelength of the source.

Insects, such as fruit flies and bees, detect near field sounds using loosely attached hairs or antennae which vibrate with air particle movement. (Tympanal organs detect the pressure component of sound.) Near-field sound, because of the rapid dissipation of energy, is suitable only for very close communication. Two examples of near-field sound communication are bee's waggle dance and Drosophila courtship songs. In fruit flies, the arista of the antennae and the third segment act as the sound receiver. Vibrations of the receiver cause rotation of the third segment, which channels sound input to the mechanoreceptors of the Johnston's Organ.

In Hawk Moths

The Johnston's organ plays a role in the control of flight stability in hawk moths. Kinematic data measured from hovering moths during steady flight indicate that the antennae vibrate with a frequency matching wingbeat (27 Hz). During complex flight, however, angular changes of the flying moth cause coriolis forces, which are predicted to manifest as a vibration of the antenna of at about twice wingbeat frequency (~60 Hz). When antennae were manipulated to vibrate at a range of frequencies and the resulting signals from the neurons associated with the Johnston's organs were measured, the response of the scolopidia neurons to the frequency was tightly coupled in the range of 50–70 Hz, which is the predicted range of vibrations caused by Coriolis effects. Thus, the Johnston's organ is tuned to detect angular changes during maneuvering in complex flight.

In Honeybees

Dancing honeybees (*Apis mellifera*) describe the location of nearby food sources by emitted airborne sound signals. These signals consist of rhythmic high-velocity movement of air particles. These near field sounds are received and interpreted using the Johnston's organ in the pedicel of the antennae. Honeybees also perceive electric field changes via the Johnston's organs in their antennae and possibly other mechano-receptors. Electric fields generated by movements of the wings cause displacements of the antennae based on Coulomb's law. Neurons of the Johnston's organ respond to movements within the range of displacements caused by electric fields. When the antennae were prevented from moving at the joints containing the Johnston's organ, bees no longer responded to biologically relevant electric fields. Honeybees respond differently to different temporal patterns. Honeybees appear to use the electric field emanating from the dancing bee for distance communication.

Mauthner Cell

The Mauthner cells are a pair of big and easily identifiable neurons (one for each half of the body) located in the rhombomere 4 of the hindbrain in fish and amphibians that are responsible for a very fast escape reflex (in the majority of animals – a so-called C-start response). The cells are also notable for their unusual use of both chemical and electrical synapses.

Evolutionary History

Mauthner cells first appear in lampreys (being absent in hagfish and lancelets), and are present in virtually all teleost fish, as well as in amphibians (including postmetamorphic frogs and toads). Some fish, such as lumpsuckers, seem to have lost the Mauthner cells however.

Role in Behavior

The C-start

A C-start is a type of a very quick startle or escape reflex that is employed by fish and amphibians (including larval frogs and toads). There are two sequential stages in the C-start: first, the head rotates about the center of mass towards the direction of future escape, and the body of the animal exhibits a curvature that resembles a letter C; then, at the second stage, the animal is propelled forward. The duration of these stages varies from species to species from about 10 to 20 ms for the first stage, and from 20 to 30 ms for the second. In fish this forward propulsion does not require contraction of the antagonistic muscle, but results from the body stiffness and the hydrodynamic resistance of the tail. When an antagonistic muscular contraction does occur during stage 2, the fish rotates in the opposite direction, producing a counter-turn, and a directional change.

The Role of the Mauthner Cell In The C-start Behavior

In cases when an abrupt acoustic, tactile or visual stimulus elicits a single action potential in one M-cell, it always correlates with a contralateral C-start escape. An extremely quick mutual feedback inhibitory circuit then assures that only one M-cell reaches spiking threshold—as the C-start has to be unilateral by definition—and that only one action potential is fired.

The Mauthner cell-mediated C-start reflex is very quick, with about 5-10 ms latency between the acoustic/tactile stimulus and the Mauthner cell discharge, and only about 2 ms between the discharge and the unilateral muscle contraction. Mauthner cells are thus the quickest motor neuron to respond to the stimulus. It makes the C-start response behaviorally important as a way to initiate the escape reflex in an *all or nothing* fashion, while the direction and speed of the escape can be corrected later through the activity of smaller motor neurons.

In larval zebrafish about ~60% of the total population of reticulospinal neurons are also activated by a stimulus that elicits the M-spike and C-start escape. A well-studied group of these reticulospinal neurons are the bilaterally paired M-cell homologues denoted MiD2cm and MiD3cm. These neurons exhibit morphological similarities to the M-cell including a lateral and ventral dendrite. They are located in rhombomeres 5 and 6 of hindbrain respectively, and also receive auditory input in parallel with the M-cell from the pVIIIth nerve. In fish, water jet stimuli that activate these neurons elicit non-mauthner initiated C-starts of a longer latency, compared with M-cell associated ones.

Although the M-cell is often considered the prototype of a command neuron in vertebrates, this designation may not be fully warranted. Although electrical stimulation of the M-cell is sufficient for eliciting a C-start, this C-start is normally weaker than the one evoked by a sensory stimulus. Moreover, the C-start can be evoked even with the M-cell ablated, although in this case the latency of the response increases. The most widely accepted model of the M-cell system, or brainstem escape network, is that the M-cell initiates a fixed action pattern to the left or right by activating a spinal motor circuit initially described by J. Diamond and colleagues, but the precise trajectory of the escape is encoded by population activity in the other classes of reticulospinal neurons functioning in parallel to the M-cell. This notion is supported by studies using *in vivo* calcium imaging in larval zebrafish which show that MiD2cm and MiD3cm are activated along with the M-cell when an offending stimulus is directed towards the head but not the tail, and are correlated with C-starts of a larger initial turn angle.

Another component of the escape response is mediated by cranial relay neurons that are activated by the Mauthner cell spike. These neurons are electrically coupled with motoneurons which innervate extraocular, jaw and opercular muscles and mediate pectoral fin adduction in hatchetfish. This component of the neural circuit was first described by Michael V.L. Bennett and colleagues.

Mauthner Cells in other Types of Behavior

Mauthner cells may be involved into behavioral patterns other than the C-start, if these types of behavior also require extremely quick bending movement of the body. Thus in goldfish Mauthner cells are activated during prey capture near the surface of the water, as this type of hunting is dangerous for the fish, and it would benefit from leaving the surface as soon as possible after the prey is captured.

In adult postmetamorphic anurans (frogs and toads) that do not have a tail, M-cells are nevertheless preserved and their discharges are associated with rapid movement of legs during an escape.

Morphology and Connections

Inputs to the M-cell: Excitation and Feed Forward Inhibition

The M-cell has two primary aspiny (lacking dendritic spines) dendrites which receive segregated inputs from various parts of the neural system. One dendrite projects laterally and the other projects either in the ventral or medial direction, depending on the species.

The ventral dendrite receives information from the optic tectum and spinal cord while the lateral dendrite receives inputs from the octovolateralis systems (the lateral line, acoustic inputs from the inner ear, and inertial information from the statoliths brought by the cranial nerve VIII).

The fibers from the ipsilateral cranial nerve VIII terminate in excitatory mixed electrical and glutamatergic synapses on the M-cell. They also electrically activate glycinergic inhibitory interneurons that terminate on the M-cells. Despite the inhibitory input having one more synapse in its pathway, there is no delay between the excitation and inhibition because the intervening synapse is electrical. It was shown that for weak stimuli the inhibition wins over the excitation, preventing the M-cell from a discharge, while for stronger stimuli excitation becomes dominant. The Inner ear afferents also terminate with electrical synapses on a population PHP inhibitory interneurons (see below) to provide an additional level of feed forward inhibition. The Mauthner cell also has GABA-, dopamine-, serotonin- and somatostatinergic inputs, each restricted to certain dendritic region.

Inputs from the optic tectum and the lateral line help control which way the C-startle bends by biasing the mauthner cells when there are obstacles in the vicinity. In cases where movement away from the stimulus is blocked, the fish may bend towards the disturbance.

Axon Cap

The Mauthner cell axon hillock is surrounded by a dense formation of neuropil, called the axon cap. The high resistance of this axon cap contributes to the typical shape of the Mauthner cell field potential. In its most advanced form the axon cap consists of a core, immediately adjacent to the Mauthner cell axon, and containing a network of very thin unmyelinated fibers, and a peripheral part. This peripheral part contains the large unmyelinated fibers of the PHP neurons that mediate the inhibitory feedback to the Mauthner cell; the Mauthner cell itself also sends small dendrites from its axon hill to the peripheral part of the axon cap. Finally, the surface of the axon cap is covered with a *cap wall* composed of several layers of astrocyte-like glial cells. Both glial cells and the unmyelinated fibers are coupled with each other by means of gap junctions.

Evolutionarily, the axon cap is a more recent development than the Mauthner cell itself, so some animals, such as lampreys and eels, while having functional Mauthner cells, don't have axon cap at all, while some other animals, such as amphibia and lungfish, do have a very simplified version of it.

Feedback Network

The main part of the Mauthner cell-associated network is the negative feedback network, which assures that only one of the two Mauthner cells fires in response to the stimulus and that, whichever Mauthner cell fires, it does so only once. Both these requirements are quite natural considering that the consequences of a single Mauthner cell discharge are so strong; a failure to comply with these two rules would not only prevent the animal from escaping, but could even physically damage it. The fastest part of this negative feedback network, which is also the one closest to the Mauthner cell, is that of the so-called *passive hyperpolarizing field potential* or PHP neurons. The fibers of these neurons are located in the axon cap, and they receive inputs from both ipsilateral and contralateral Mauthner cells. The field potentials of PHP neurons are strongly positive, and form a part of the 'Signature field potential' of the Mauthner cell, with the early (ipsilaterally initiated) component being called the Extracellular Hyperpolarizing Potential (EHP), and the later (contralaterally initiated) component being sometimes addressed in the literature as the Late Collateral Inhibition (LCI). The action of PHP neurons onto the Mauthner cells is mediated by electrical, and not chemical effects: the outward currents generated by the action potentials in axon cap fibers flow inward across the Mauthner cell axon hillock and hyperpolarize it.

Outputs

The only axon of the Mauthner cell reaches from the cell to the midline of the hindbrain, promptly crosses it to the contralateral side, and then descends caudally along the spinal cord. A single discharge of the M-cell achieves a whole set of parallel effects onto the spinal motor networks: 1) it monosynaptically excites large primary motoneurons at one side of the body; 2) disinaptically excites smaller motoneurons at the same side of the body; 3) initiates action potentials in inhibitory interneurons electrically coupled to the M-cell axon, and by their means inhibits a) inhibitory interneurons still at the same side of the body (to prevent them from interfering with the C-start), as well as b) motoneurons at the other side of the body. As a result of this pattern of activation the quick muscles at one side of the body contract simultaneously, while the muscles at the other side of the body relax.

Electrophysiology

Ephaptic Properties

The inhibition of the M-cell by the PHP cells occurs by ephaptic interactions. The inhibition is brought about without a chemical synapses or electrical synaptic coupling having low resistance gap junctions joining the cells. When the region of the PHP cell axon outside the axon cap depolarizes, the influx of positive charge into the cell through voltage gated sodium channels is accompanied by a passive outflow of current from the PHP cell axon into the region bound by the axon cap. Due to the high resistance of the surrounding glial cells, the charge does not dissipate and the potential across the M-cell membrane is increased, hyperpolarizing it.

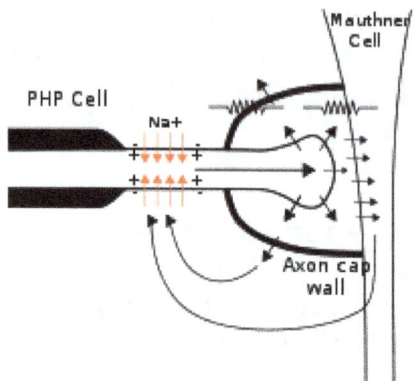

Ephaptic inhibition at the mauthner axon cap by PHP cells

Signature Field Potential

Because of its size, presence of a quick feedback network, and abundance of electrical and qua-si-electrical (ephaptic) synapses, the Mauthner cell has a strong field potential of a very characteristic shape. This field potential starts with a high-amplitude potential sink up to tens of millivolts in amplitude that originates from the Mauthner cell discharge, and which is closely followed by a positive potential, called Extrinsic Hyperpolarizing Potential or EHP, which is associated with the activity of the recurrent feedback network.

Due to its high amplitude, in some animals the negative part of Mauthner cell field potential can be detected up to several hundred micrometres away from the cell itself. The positive components of the field potential are strongest in the axon cap, reaching amplitudes of 45 mV in adult goldfish. With a knowledge of these properties of the field potential, it is possible to use field potential monitoring as a way to find the Mauthner cell body in vivo, or in vitro in a whole brain preparation, moving the recording electrode in the hindbrain, while at the same time stimulating the spinal cord, thus evoking antidromic action potentials in the Mauthner cell axon.

Plasticity

Application of serotonin was shown to increase inhibitory inputs to the M-cell, while application of dopamine – to increase the amplitude of both chemical and electrical components of the VIIIth nerve responses via a G protein-mediated activation of postsynaptic D2 receptor. An activity-dependent LTP can be evoked in M-cells by a high-frequency stimulation of the VIIIth nerve. Surprisingly, this LTP is electrical synapse-mediated, and is presumed to involve modification of the gap junction channels. A possibility of LTP induction by sensory stimuli in vivo, and the evidence for the LTP of inhibitory inputs to M-cells were also demonstrated.

Spontaneous preference in turn direction in young goldfish is correlated with one of the Mauthner cells being bigger than the other one. It is possible to change the preference of fish by raising them in conditions facilitating turns in a specific direction; this shift is accompanied by a correspondent change in M-cell sizes.

History of Research

The Mauthner cell was first identified by the Viennese ophthalmologist Ludwig Mauthner in the teleost fish for its associated neural circuit which mediates an escape response called the C-start or C-startle to direct the fish away from a predator.

The M-cell is a model system in the field of Neuroethology. The M-cell system has served for detailed neurophysiological and histological investigations of synaptic transmission and synaptic plasticity. Studies by Donald Faber and Henri Korn helped to establish the one vesicle hypothesis of synaptic transmission in the CNS. Other important research topics that have been investigated in the M-cell system include studies by Yoichi Oda and colleagues on inhibitory long-term potentiation and auditory conditioning of the startle response, and studies by Alberto Pereda and colleagues on plasticity of electrical synapses. Other research topics investigated in the M-cell system include studies of spinal neural networks and neural regeneration by Joe Fetcho and colleagues, as well as underwater sound localization, and the biophysics of computation in single neurons.

Medullary Command Nucleus

The medullary command nucleus (MCN), also called the *pacemaker nucleus*, is a group of nerve cells found in the bodies of weakly electric fish. It controls the function of electrocytes by regulating the frequency of electrical impulses. Signals originating in the MCN are transmitted to electrocytes, where changes in ion concentration cause electrical charges to be generated. The nucleus both sends and receives signals, thereby acting as a regulator and central processor for the electro sensors in the fish's body. Inputs into the MCN originate in the mesencephalic precommand nucleus, thalamic dorsal posterior nucleus, and toral ventroposterior nucleus. All of these nuclei have dense projections into the MCN, with the exception of theToral Ventroposterior nucleus, which contain only a ventral edge projection.

References

- Plachta D T T; Hanke W; Bleckmann H (2003). "A hydrodynamic topographic map in the midbrain of goldfish Carassius auratus". Journal of Experimental Biology. 206 (19): 3479–86. doi:10.1242/jeb.00582

- Journal of Undergraduate Life Sciences. "Appropriate maze methodology to study learning in fish" (PDF). Retrieved 28 May 2009

- Graham, Michael (1941). "Sense of Hearing in Fishes". Nature. 147: 779. Bibcode:1941Natur.147..779G. doi:10.1038/147779b0

- Meyer CG; Holland KN; Papastamatiou YP (2005). "Sharks can detect changes in the geomagnetic field". Journal of the Royal Society, Interface. 2 (2): 129–30. PMC 1578252. PMID 16849172. doi:10.1098/rsif.2004.0021

- Bever MM, Borgens RB (January 1988). "Eye regeneration in the mystery snail". The Journal of Experimental Zoology. 245 (1): 33–42. PMID 3351443. doi:10.1002/jez.1402450106

- Popper, A. N.; Platt, C. (1993). "Inner ear and lateral line of bony fishes". In Evans, D. H. The Physiology of Fishes (1st ed.). CRC Press. pp. 99–136. ISBN 978-0-8493-8042-6

- Boehm T; Zufall F (February 2006). "MHC peptides and the sensory evaluation of genotype". Trends in Neurosciences. 29 (2): 100–7. PMID 16337283. doi:10.1016/j.tins.2005.11.006

- Cooper, W. E. 1995a. Evolution and function of lingual shape in lizards, with emphasis on elongation, extensibility, and chemical sampling. Journal of Chemical Ecology 21:477-505

- Chase R.: Sensory Organs and the Nervous System. in Barker G. M. (ed.): The biology of terrestrial molluscs.

CABI Publishing, Oxon, UK, 2001, ISBN 0-85199-318-4. 1-146, cited pages: 179-211

- "Rose, J.D. 2003. A Critique of the paper: "Do fish have nociceptors: Evidence for the evolution of a vertebrate sensory system"" (PDF). Retrieved 21 May 2011

- Hughes HP (August 1976). "Structure and regeneration of the eyes of strombid gastropods". Cell and Tissue Research. 171 (2): 259–71. PMID 975213. doi:10.1007/BF00219410

- Johnston, C (1855). "Auditory Apparatus of the Culex Mosquito" (PDF). Quarterly Journal of Microscopical Science. 3: 97–102

- Dreller, C.; Kirchner, W. H. (1993). "Hearing in honeybees: localization of the auditory sense organ". Journal of Comparative Physiology A. 173 (3): 275–279. doi:10.1007/bf00212691

- Götting, Klaus-Jürgen (1994). "Schnecken". In Becker, U.; Ganter, S.; Just, C.; Sauermost, R. Lexikon der Biologie. Heidelberg: Spektrum Akademischer Verlag. ISBN 3-86025-156-2

- Canfield JG, Rose GJ (1993). "Activation of Mauthner neurons during prey capture". Journal of Comparative Physiology A. 172 (5): 611–618. doi:10.1007/BF00213683

- Shupak A. Sharoni Z. Yanir Y. Keynan Y. Alfie Y. Halpern P. (January 2005). "Underwater Hearing and Sound Localization with and without an Air Interface". Otology & Neurotology. 26 (1): 127–130. doi:10.1097/00129492-200501000-00023

4

Bird Anatomy: An Integrated Study

Bird anatomy is the study of the internal and external parts of a bird. The subject studies all the systems related to birds. Some examples of these systems are skeletal system, muscular system, respiratory system, digestive system and circulatory system. The topics discussed in the chapter are of great importance to broaden the existing knowledge on bird anatomy.

Bird Anatomy

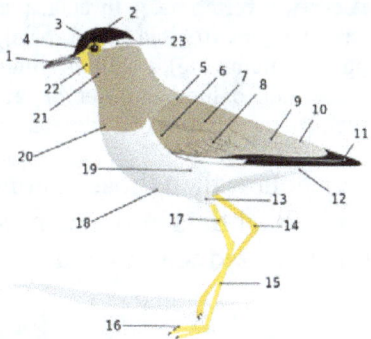

External anatomy (topography) of a typical bird: 1 Beak, 2 Head, 3 Iris, 4 Pupil, 5 Mantle, 6 Lesser coverts 7 Scapulars, 8 Coverts, 9 Tertials, 10 Rump, 11 Primaries, 12 Vent, 13 Thigh, 14 Tibio-tarsal articulation 15 Tarsus, 16 Feet, 17 Tibia, 18 Belly, 19 Flanks, 20 Breast, 21 Throat, 22 Wattle, 23 Eyestripe

Bird anatomy, or the physiological structure of birds' bodies, shows many unique adaptations, mostly aiding flight. Birds have a light skeletal system and light but powerful musculature which, along with circulatory and respiratory systems capable of very high metabolic rates and oxygen supply, permit the bird to fly. The development of a beak has led to evolution of a specially adapted digestive system. These anatomical specializations have earned birds their own class in the vertebrate phylum.

Skeletal System

The bird skeleton is highly adapted for flight. It is extremely lightweight but strong enough to withstand the stresses of taking off, flying, and landing. One key adaptation is the fusing of bones into single ossifications, such as the pygostyle. Because of this, birds usually have a smaller number of bones than other terrestrial vertebrates. Birds also lack teeth or even a true jaw, instead having a beak, which is far more lightweight. The beaks of many baby birds have a projection called an egg tooth, which facilitates their exit from the amniotic egg, and that falls off once it has done its job.

Birds have many bones that are hollow (pneumatized) with criss-crossing struts or trusses for structural strength. The number of hollow bones varies among species, though large gliding and soaring birds tend to have the most. Respiratory air sacs often form air pockets within the semi-hollow bones of the bird's skeleton.

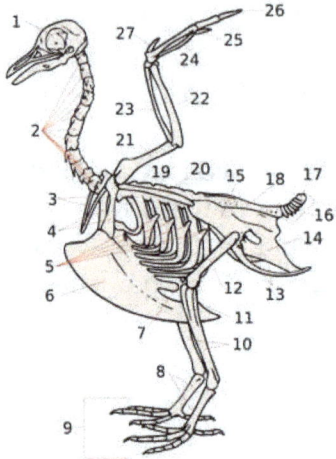

A stylised dove skeleton. Key: 1. skull, 2. cervical vertebrae, 3. furcula, 4. coracoid, 5. uncinate processes of ribs 6. keel, 7. patella, 8. tarsometatarsus, 9. digits, 10. tibia (tibiotarsus), 11. fibia (tibiotarsus), 12. femur 13. ischium (innominate), 14. pubis (innominate), 15. illium (innominate), 16. caudal vertebrae 17. pygostyle, 18. synsacrum, 19. scapula, 20. lumbar vertebrae, 21. humerus, 22. ulna, 23. radius 24. carpus (carpometacarpus), 25. metacarpus (carpometacarpus), 26. digits, 27. alula

The bones of diving birds are often less hollow than those of non-diving species. Loons and puffins are without pneumatized bones entirely. Flightless birds, such as ostriches and emus, demonstrate osseous pneumaticity, possessing pneumatized femurs and, in the case of the emu, pneumatized cervical vertebrae.

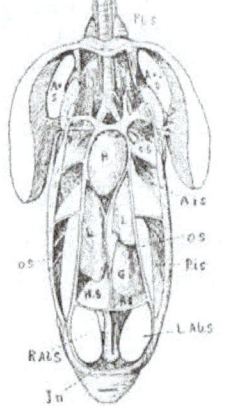

DISSECTION SHOWING THE LUNGS AND AIR-SACS OF A BIRD

Pb. s. = Pre-bronchial sac. Ax. = Axillary sac bounded externally by the breast-muscles, seen here in section. S. s. = Partition dividing anterior intermediate sac (A.i.s.) from the sub-bronchial sac. P.i.s. = Posterior intermediate sac. O.s. = Oblique septum. H.s. = Horizontal septum. L. Ab. s. = Left abdominal sac. H. = Heart. G. = Gizzard. L. = Liver. In. = Intestine. (After Strasser.)

Air-sacs and their distribution

Birds also have more cervical (neck) vertebrae than many other animals; most have a highly flexible neck consisting of 13-25 vertebrae. Birds are the only vertebrates to have fused clavicles (collarbone) (the furcula or wishbone) or a keeled sternum or breastbone. The keel of the sternum serves as an attachment site for the muscles used for flight or, similarly, for swimming, in penguins. Again, flightless birds, such as ostriches, which do not have highly developed pectoral muscles, lack a pronounced keel on the sternum. Swimming birds have a wide sternum, while walking birds have a long or high sternum and flying birds have a sternum width and height that are nearly equal.

Birds have uncinate processes on the ribs. These are hooked extensions of bone which help to strengthen the rib cage by overlapping with the rib behind them. This feature is also found in the tuatara (*Sphenodon*). They also have a greatly elongate tetradiate pelvis, similar to some reptiles. The hind limb has an intra-tarsal joint found also in some reptiles. There is extensive fusion of the trunk vertebrae as well as fusion with the pectoral girdle. They have a diapsid skull, as in reptiles, with a pre-lachrymal fossa (present in some reptiles). The skull has a single occipital condyle.

The skull consists of five major bones: the frontal (top of head), parietal (back of head), premaxillary and nasal (top beak), and the mandible (bottom beak). The skull of a normal bird usually weighs about 1% of the bird's total body weight. The eye occupies a considerable amount of the skull and is surrounded by a sclerotic eye-ring, a ring of tiny bones. This characteristic is also seen in reptiles.

The vertebral column consists of vertebrae, and is divided into three sections: cervical (11-25) (neck), synsacrum (fused vertebrae of the back, also fused to the hips (pelvis)), and pygostyle (tail).

The chest consists of the furcula (wishbone) and coracoid (collar bone), which, together with the scapula, form the pectoral girdle. The side of the chest is formed by the ribs, which meet at the sternum (mid-line of the chest).

The shoulder consists of the scapula (shoulder blade), coracoid, and humerus (upper arm). The humerus joins the radius and ulna (forearm) to form the elbow. The carpus and metacarpus form the "wrist" and "hand" of the bird, and the digits are fused together. The bones in the wing are extremely light so that the bird can fly more easily.

The hips consist of the pelvis, which includes three major bones: the ilium (top of the hip), ischium (sides of hip), and pubis (front of the hip). These are fused into one (the innominate bone). Innominate bones are evolutionary significant in that they allow birds to lay eggs. They meet at the acetabulum (hip socket) and articulate with the femur, which is the first bone of the hind limb.

The upper leg consists of the femur. At the knee joint, the femur connects to the tibiotarsus (shin) and fibula (side of lower leg). The tarsometatarsus forms the upper part of the foot, digits make up the toes. The leg bones of birds are the heaviest, contributing to a low center of gravity, which aids in flight. A bird's skeleton accounts for only about 5% of its total body weight

Feet

Birds' feet are classified as anisodactyl, zygodactyl, heterodactyl, syndactyl or pamprodactyl. Anisodactyl is the most common arrangement of digits in birds, with three toes forward and one back. This is common in songbirds and other perching birds, as well as hunting birds like eagles, hawks, and falcons.

Syndactyly, as it occurs in birds, is like anisodactyly, except that the third and fourth toes (the outer and middle forward-pointing toes), or three toes, are fused together, as in the belted kingfisher *Ceryle alcyon*. This is characteristic of Coraciiformes (kingfishers, bee-eaters, rollers, etc.).

The zygodactyly (from Greek yoke) is an arrangement of digits in birds, with two toes facing forward (digits two and three) and two back (digits one and four). This arrangement is most common in arboreal species, particularly those that climb tree trunks or clamber through foliage. Zygodactyly occurs in the parrots, woodpeckers (including flickers), cuckoos (including roadrunners), and some owls. Zygodactyl tracks have been found dating to 120–110 Ma (early Cretaceous), 50 million years before the first identified zygodactyl fossils.

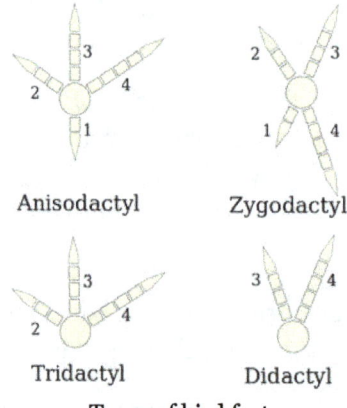

Types of bird feet

Heterodactyly is like zygodactyly, except that digits three and four point forward and digits one and two point back. This is found only in trogons, while pamprodactyl is an arrangement in which all four toes may point forward, or birds may rotate the outer two toes backward. It is a characteristic of swifts (Apodidae).

Muscular System

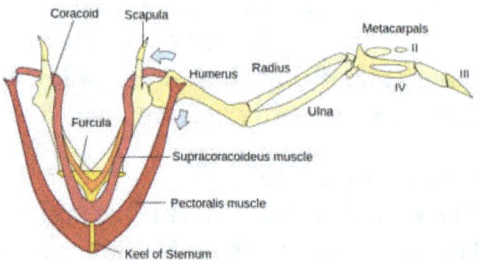

The supracoracoideus works using a pulley like system to lift the wing while the pectorals provide the powerful downstroke

Most birds have approximately 175 different muscles, mainly controlling the wings, skin, and legs. The largest muscles in the bird are the pectorals, or the breast muscles, which control the wings and make up about 15 - 25% of a flighted bird's body weight. They provide the powerful wing stroke essential for flight. The muscle medial to (underneath) the pectorals is the supracoracoideus. It raises the wing between wingbeats. Both muscle groups attach to the keel of the sternum. This is remarkable, because other vertebrates have the muscles to raise the upper limbs generally attached to areas on the back of the spine. The supracoracoideus and the pectorals together make up about 25 – 35% of the bird's full body weight.

The skin muscles help a bird in its flight by adjusting the feathers, which are attached to the skin muscle and help the bird in its flight maneuvers.

There are only a few muscles in the trunk and the tail, but they are very strong and are essential for the bird. The pygostyle controls all the movement in the tail and controls the feathers in the tail. This gives the tail a larger surface area which helps keep the bird in the air.

Integumentary System

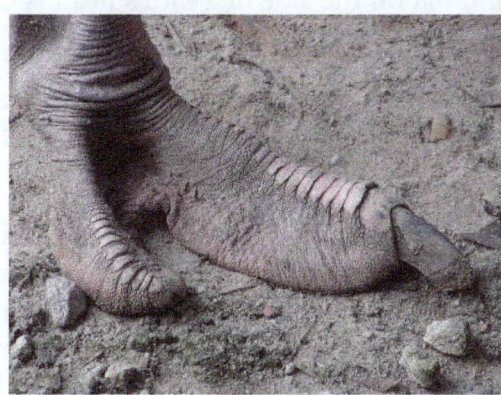

Ostrich foot integument (podotheca)

Scales

The scales of birds are composed of keratin, like beaks, claws, and spurs. They are found mainly on the toes and metatarsus, but may be found further up on the ankle in some birds. Most bird scales do not overlap significantly, except in the cases of kingfishers and woodpeckers. The scales and scutes of birds were originally thought to be homologous to those of reptiles and mammals; however, more recent research suggests that scales in birds re-evolved after the evolution of feathers.

Bird embryos begin development with smooth skin. On the feet, the corneum, or outermost layer, of this skin may keratinize, thicken and form scales. These scales can be organized into:

- Cancella – minute scales which are really just a thickening and hardening of the skin, crisscrossed with shallow grooves.

- Scutella – scales that are not quite as large as scutes, such as those found on the caudal, or hind part, of the chicken metatarsus.

- Scutes – the largest scales, usually on the anterior surface of the metatarsus and dorsal surface of the toes.

The rows of scutes on the anterior of the metatarsus can be called an "acrometatarsium" or "acrotarsium".

Reticula are located on the lateral and medial surfaces (sides) of the foot and were originally thought to be separate scales. However, histological and evolutionary developmental work in this area revealed that these structures lack beta-keratin (a hallmark of reptilian scales) and are entirely composed of alpha-keratin. This, along with their unique structure, has led to the suggestion that these are actually feather buds that were arrested early in development.

Rhamphotheca and Podotheca

The bills of many waders have Herbst corpuscles which help them find prey hidden under wet sand, by detecting minute pressure differences in the water. All extant birds can move the parts of the upper jaw relative to the brain case. However this is more prominent in some birds and can be readily detected in parrots.

The region between the eye and bill on the side of a bird's head is called the lore. This region is sometimes featherless, and the skin may be tinted, as in many species of the cormorant family.

The scaly covering present on the foot of the birds is called podotheca.

Beak

The beak, bill, or rostrum is an external anatomical structure of birds which is used for eating and for grooming, manipulating objects, killing prey, fighting, probing for food, courtship and feeding young. Although beaks vary significantly in size, shape and color, they share a similar underlying structure. Two bony projections—the upper and lower mandibles—covered with a thin keratinized layer of epidermis known as the rhamphotheca. In most species, two holes known as nares lead to the respiratory system.

Respiratory System

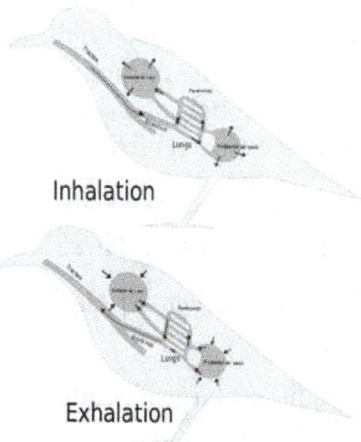

Inhalation

Exhalation

Inhalation-exhalation cycle in birds.

Due to their high metabolic rate required for flight, birds have a high oxygen demand. Their highly effective respiratory system helps them meet that demand. Although birds have lungs, which are fairly rigid structures, they rely on air sacs distributed throughout much of their bodies for ventilation, or the bellows action for inhalation and exhalation. While bird lungs are smaller in comparison to mammals, the air sacs account for 15% of the total body volume, compared to the 7% devoted to the alveoli which act as the bellows in mammals.

The walls of these sacs do not have a good blood supply and so do not play a direct role in gas exchange. They act like a set of bellows to move air unidirectionally through the parabronchi of the rigid lungs. Birds lack a diaphragm, so rather than the regular expansion and contraction of the lungs as is seen in mammals, the expansion and contraction of the air sacs allow the volume of the lungs to remain constant with oxygenated air constantly flowing in a single direction through them. The active phase of respiration in birds is exhalation, requiring muscular contraction.

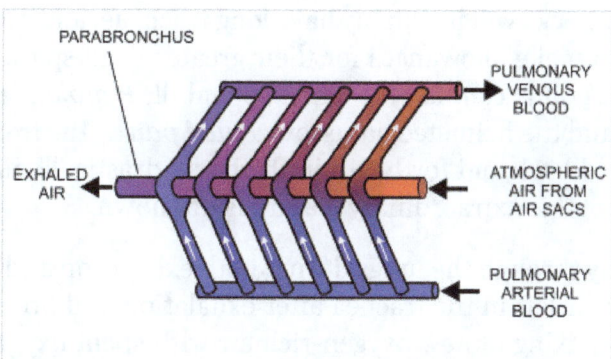

A diagrammatic representation of the cross-current respiratory gas exchanger in the lungs of birds. Air is forced from the air sacs unidirectionally (from right to left in the diagram) through the parabronchi. The pulmonary capillaries surround the parabronchi in the manner shown (blood flowing from below the parabronchus to above it in the diagram). Blood or air with a high oxygen content is shown in red; oxygen-poor air or blood is shown in various shades of purple-blue.

Three distinct sets of organs perform respiration — the anterior air sacs (interclavicular, cervicals, and anterior thoracics), the lungs, and the posterior air sacs (posterior thoracics and abdominals). Typically there are nine air sacs within the system; however, that number can range between seven and twelve, depending on the species of bird. Passeriformes possess seven air sacs, as the clavicular air sacs may interconnect or be fused with the cranial thoracic sacs.

During inhalation, atmospheric air initially enters the bird through the nares (nostrils) where it is heated, humidified, and filtered. From there, the air enters the trachea and continues beyond the syrinx at which point the trachea branches into two primary bronchi. The primary bronchi enter the lungs to become the intrapulmonary bronchi, which give off a set of parallel branches called ventrobronchi and, a little further on, an equivalent set of dorsobronchi. The ends of the intrapulmonary bronchi discharge air into the posterior air sacs at the caudal end of the bird. Each pair of dorso-ventrobronchi are connected by a large number of parallel microscopic paleoparabronchi (commonly referred to as parabronchi) where gas exchange occurs. As the bird inhales, tracheal air flows through the intrapulmonary bronchi into the posterior air sacs, as well as into the *dorso*bronchi (but not into the ventrobronchi whose entrances from the intrapulmonary bronchi are tightly closed during inhalation). From the dorsobronchi the air flows through the parabronchi (and therefore the gas exchanger) to the ventrobronchi from where the air can only escape into the expanding anterior air sacs. So, during inhalation, both the posterior and anterior air sacs expand.

During exhalation the intrapulmonary bronchi constrict (i.e. close up) between the region where the ventrobronchi branch off and the region where the dorsobronchi branch off. The contracting posterior air sacs can therefore only empty into the dorsobronchi. From there the fresh air from the posterior air sacs flows through the parabronchi (in the same direction as occurred during inhalation) into ventrobronchi. The air passages connecting the ventrobronchi and anterior air sacs to the intrapulmonary bronchi open up during exhalation, thus allowing oxygen-poor air from these two organs to escape via the trachea to the exterior.

The trachea is an area of dead space: the oxygen-poor air it contains at the end of exhalation is the first air to re-enter the posterior air sacs and lungs. In comparison to a mammalian respiratory tract, the dead space volume in a bird is, on average, 4.5 times greater than in mammals of the same size. Birds with long necks will inevitably have long tracheae, and must therefore take deeper breaths than mammals to make allowances for their greater dead space volumes. In some birds (e.g. the whooper swan, *Cygnus cygnus*, the white spoonbill, *Platalea leucorodia*, the whooping crane, *Grus americana*, and the helmeted curassow, *Pauxi pauxi*) the trachea, which some cranes can be 1.5 m long, is coiled back and forth within the body, drastically increasing the dead space ventilation. The purpose of this extraordinary feature is unknown.

Air passes unidirectionally through the lungs during both exhalation and inspiration, causing, except for the oxygen-poor air left in the trachea after exhalation and breathed in at the beginning of inhalation, little to no mixing of new oxygen-rich air with spent oxygen-poor air (as occurs in mammalian lungs), changing only (from oxygen-rich to oxygen-poor) as it moves (unidirectionally) through the parabronchi.

Avian lungs do not have alveoli as mammalian lungs do, but instead contain millions of narrow passages known as parabronchi, connecting the dorsobronchi to the ventrobronchi at either ends of the lungs. Air flows anteriorly (caudal to cranial) through the parallel, honeycombed walls of the parabronchi. The cells of the honeycomb are dead-end air vesicles, called *atria*, which project radially from the parabronchi. The *atria* are the site of gas exchange by simple diffusion. The blood flow around the parabronchi (and their atria), forms a cross-current gas exchanger.

The partial pressure of oxygen in the parabronchi declines along their length as O_2 diffuses into the blood. The capillaries leaving the exchanger near the entrance of airflow take up more O_2 than the capillaries leaving near the exit end of the parabronchi. When the contents of all capillaries mix, the final partial pressure of oxygen of the mixed pulmonary venous blood is higher than that of the exhaled air, but is nevertheless less half that of the inhaled air, thus achieving roughly the same systemic arterial blood partial pressure of oxygen as mammals do with their bellows-type lungs.

All species of birds with the exception of the penguin, have a small region of their lungs devoted to "neopulmonic parabronchi". This unorganized network of microscopic tubes branches off from the posterior air sacs, and open haphazardly into both the dorso- and ventrobronchi, as well as directly into the intrapulmonary bronchi. Unlike the paleoparabronchi, in which the air moves unidirectionally, the air flow in the neopulmonic parabronchi is bidirectional. The neopulmonic parabronchi never make up more than 25% of the total gas exchange surface of birds.

The syrinx is the sound-producing vocal organ of birds, located at the base of a bird's trachea. As with the mammalian larynx, sound is produced by the vibration of air flowing across the organ. The syrinx enables some species of birds to produce extremely complex vocalizations, even mimicking human speech. In some songbirds, the syrinx can produce more than one sound at a time.

Circulatory System

Birds have a four-chambered heart, in common with humans, most mammals, and some reptiles (mainly the crocodilia). This adaptation allows for an efficient nutrient and oxygen transport throughout the body, providing birds with energy to fly and maintain high levels of activity. A ruby-throated hummingbird's heart beats up to 1200 times per minute (about 20 beats per second).

The human heart (left) and chicken heart (right) share many similar characteristics. Avian hearts pump faster than mammalian hearts. Due to the faster heart rate, the muscles surrounding the ventricles of the chicken heart are thicker. Both hearts are labeled with the following parts: 1. Ascending Aorta 2. Left Atrium 3. Left Ventricle 4. Right Ventricle 5. Right Atrium. In chickens and others birds, the superior cava is double.

Digestive System

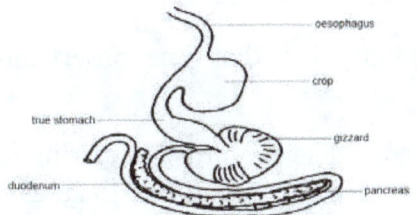

Simplified depiction of avian digestive system.

Many birds possess a muscular pouch along the esophagus called a crop. The crop functions to both soften food and regulate its flow through the system by storing it temporarily. The size and shape of the crop is quite variable among the birds. Members of the order Columbiformes, such as pigeons, produce a nutritious crop milk which is fed to their young by regurgitation. The avian stomach is composed of two organs that work together during digestion: The Proventriculus and the Gizzard. The Proventriculus is a rod shaped tube, which is found between the Oesophagus (Connection between throat and stomach) and the Ventriculus, that secretes hydrochloric acid and Pepsinogen into the digestive tract. Pepsinogen produces pepsin, which breaks the peptide bonds found in amino acids. These gastric juices are then mixed with the stomach content through the muscular and mechanical (gastroliths) mechanisms of the Gizzard. The *ventriculus*, or gizzard, is composed of four muscular bands that rotate and crush food by shifting the food from one area to the next within the gizzard. The gizzard of some species contains small pieces of grit or stone swallowed by the bird to aid in the grinding process of digestion, serving the function of mammalian or reptilian teeth. The use of gizzard stones is a similarity between birds and dinosaurs, which left gizzard stones called gastroliths as trace fossils.

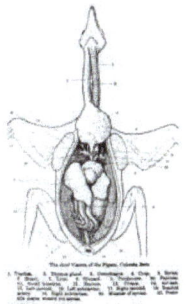

Alimentary canal of the bird exposed

Drinking Behavior

There are four general ways in which birds drink: using gravity itself, sucking, use of the tongue, and deriving water entirely from food.

Most birds are unable to swallow by the "sucking" or "pumping" action of peristalsis in their esophagus (as humans do), and drink by repeatedly raising their heads after filling their mouths to allow the liquid to flow by gravity, a method usually described as "sipping" or "tipping up". The notable exception is the Columbiformes; in fact, according to Konrad Lorenz in 1939:

One recognizes the order by the single behavioral characteristic, namely that in drinking the water is pumped up by peristalsis of the esophagus which occurs without exception within the order. The only other group, however, which shows the same behavior, the Pteroclidae, is placed near the doves just by this doubtlessly very old characteristic.

Although this general rule still stands, since that time, observations have been made of a few exceptions in both directions.

In addition, specialized nectar feeders like sunbirds (Nectariniidae) and hummingbirds (Trochilidae) drink by using protrusible grooved or trough-like tongues, and parrots (Psittacidae) lap up water.

Many seabirds have glands near the eyes that allow them to drink seawater. Excess salt is eliminated from the nostrils. Many desert birds get the water that they need entirely from their food. The elimination of nitrogenous wastes as uric acid reduces the physiological demand for water.

Reproductive and Urogenital Systems

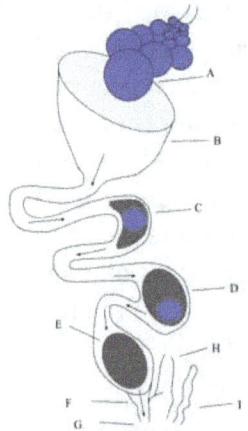

A diagram of a female chicken reproduction system. A .Mature ovum, B. Infundibulum, C. Magnum, D. Isthmus, E. Uterus, F. Vagina, G. Cloaca, H. Large intestine, I. rudiment of right oviduct

Male birds have two testes which become hundreds of times larger during the breeding season to produce sperm. The testes in male birds are generally asymmetric with most birds having a larger left testis. Female birds in most families have only one functional ovary (the left one), connected to an oviduct — although two ovaries are present in the embryonic stage of each female bird. Some species of birds have two functional ovaries, and the order Apterygiformes always retain both ovaries.

Fledgling

Most male birds have no phallus. In the males of species without a phallus, sperm is stored in the seminal glomera within the cloacal protuberance prior to copulation. During copulation, the female moves her tail to the side and the male either mounts the female from behind or in front (as in the stitchbird), or moves very close to her. The cloacae then touch, so that the sperm can enter the female's reproductive tract. This can happen very fast, sometimes in less than half a second.

The sperm is stored in the female's sperm storage tubules for a period varying from a week to more than 100 days, depending on the species. Then, eggs will be fertilized individually as they leave the ovaries, before the shell is calcified in the oviduct. After the egg is laid by the female, the embryo continues to develop in the egg outside the female body.

A juvenile laughing gull

Many waterfowl and some other birds, such as the ostrich and turkey, possess a phallus. This appears to be the primitive condition among birds, most birds have lost the phallus. The length is thought to be related to sperm competition in species that usually mate many times in a breeding season; sperm deposited closer to the ovaries is more likely to achieve fertilization. The longer and more complicated phalli tend to occur in waterfowl whose females have unusual anatomical features of the vagina (such as dead end sacs and clockwise coils). These vaginal structures may be used to prevent penetration by the male phallus (which coils counter-clockwise). In these species, copulation is often violent and female co-operation is not required; the female ability to prevent fertilization may allow the female to choose the father for her offspring. When not copulating, the phallus is hidden within the proctodeum compartment within the cloaca, just inside the vent.

After the eggs hatch, parents provide varying degrees of care in terms of food and protection. Precocial birds can care for themselves independently within minutes of hatching; altricial hatchlings are helpless, blind, and naked, and require extended parental care. The chicks of many ground-nesting birds such as partridges and waders are often able to run virtually immediately after hatching; such birds are referred to as nidifugous. The young of hole-nesters, on the other hand, are often totally incapable of unassisted survival. The process whereby a chick acquires feathers until it can fly is called "fledging".

Some birds, such as pigeons, geese, and red-crowned cranes, remain with their mates for life and may produce offspring on a regular basis.

Kidney

Avian kidneys function in almost the same way as the more extensively studied mammalian kidney, but with a few important adaptations; while much of the anatomy remains unchanged in design, some important modifications have occurred during their evolution. A bird has paired kidneys which are connected to the lower gastrointestinal tract through the ureters. Depending on the bird species, the cortex makes up around 71-80% of the kidney's mass, while the medulla is much smaller at about 5-15% of the mass. Blood vessels and other tubes make up the remaining mass. Unique to birds is the presence of two different types of nephrons (the functional unit of the kidney) both reptilian-like nephrons located in the cortex and mammalian-like nephrons located in the medulla. Reptilian nephrons are more abundant but lack the distinctive loops of Henle seen in mammals. The urine collected by the kidney is emptied into the cloaca through the ureters and then to the colon by reverse peristalsis.

Nervous System

Birds have acute eyesight—raptors have vision eight times sharper than humans—thanks to higher densities of photoreceptors in the retina (up to 1,000,000 per square mm in *Buteos*, compared to 200,000 for humans), a high number of neurons in the optic nerves, a second set of eye muscles not found in other animals, and, in some cases, an indented fovea which magnifies the central part of the visual field. Many species, including hummingbirds and albatrosses, have two foveas in each eye. Many birds can detect polarised light.

The avian ear is adapted to pick up on slight and rapid changes of pitch found in bird song. General avian tympanic membrane form is ovular and slightly conical. Morphological differences in the middle ear are observed between species. Ossicles within green finches, blackbirds, song thrushes, and house sparrows are proportionately shorter to those found in pheasants, Mallard ducks, and sea birds. In song birds, a syrinx allows the respective possessors to create intricate melodies and tones. The middle avian ear is made up of three semicircular canals, each ending in an ampulla and joining to connect with the macula sacculus and lagena, of which the cochlea, a straight short tube to the external ear, branches from.

Birds have a large brain to body mass ratio. This is reflected in the advanced and complex bird intelligence.

Immune System

The immune system of birds resembles that of other animals. Birds have both innate and adaptive immune systems. Birds are susceptible to tumours, immune deficiency and autoimmune diseases.

Bursa of Fabricius

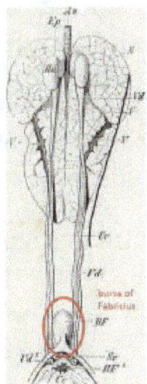

Internal view of the location of bursa of fabricius

Function

The bursa of fabricius, also known as the cloacal bursa, is a lymphoid organ which aids in the production of B lymphocyte during Humoral immunity. The bursa of fabricius is present during juvenile stages but curls up and is not visible after the sparrow reaches sexual maturity.

Anatomy

The bursa of fabricius is a circular pouch connected to the superior dorsal side of the cloaca . The bursa is composed of many folds, known as plica, which are lined by more than 10,000 follicles encompassed by connective tissue and surrounded by mesenchyme. Each follicle consists of a cortex that surrounds a medulla. The cortex houses the highly compacted B lymphocytes, whereas the medulla houses lymphocytes loosely. The medulla is separated from the lumen by the epithelium and this aids in the transport of epithelial cells into the lumen of the bursa. There are 150,000 B lymphocytes located around each follicle.

Comb (Anatomy)

A rooster with a large red comb

A comb is a fleshy growth or crest on the top of the head of gallinaceous birds, such as turkeys, pheasants, and domestic chickens. Its alternative name cockscomb (with several spelling variations) reflects that combs are generally larger on males than on females (a male gallinaceous bird is called a cock). There can be several fleshy protuberances on the heads and throats of gallinaceous birds, i.e. comb, wattle, ear lobes and nodules, which collectively are called caruncles, however, in turkeys caruncle refers specifically to the fleshy nodules on the head and throat.

Chicken combs are most commonly red (but may be black or dark purple in breeds such as Silkies or Sebrights), but in other species the color may vary from light grey to deep blue or red; turkey combs can vary in color from bright red to blue.

The comb may be a reliable indicator of health or vigor and is used for mate-assessment in some poultry species.

Shape

Comb shape varies considerably depending on the breed or species of bird. The "comb" most often refers to chickens in which the most common shape is the "single comb" of a rooster from breeds such as the Leghorn. Other common comb types are the "rose comb" (e.g. the Rosecomb) or "pea comb" (e.g. the Brahma or the Araucana). Other distinctive shapes have been selectively bred for, such as the "buttercup comb" of the Sicilian Buttercup, "V combs" (popularly called "devil horn") in the Houdan and other breeds, the "cushion comb" of the Chantecler, and "walnut comb" of Malay game.

In Cookery

Combs are used in cookery, often in combination with wattles or chicken kidneys.

Combs were formerly used in French cuisine as garnishes. They were also used to prepare salpi-cons served in vol-au-vents, profiteroles, etc. in which they were often combined with other luxury ingredients such as truffles, sweetbreads, or morels in a cream sauce.

In Italian cuisine, combs are an important ingredient in the famous sauce called *Cibreo*, which also includes chicken livers, wattles, and unlaid eggs. It is used as a sauce for tagliatelle and in the molded potato-ricotta ring *Cimabella con cibreo*.

Combs are prepared by parboiling and skinning, then cooking in court-bouillon. After preparation, they are greyish.

Rooster combs are often served in Chinese dim sum style dishes.

Other

Because of its bright color and distinctive shape, "cockscomb" also describes various plants, including the florists' plant *Celosia cristata*, the meadow weed yellow rattle, sainfoin, wild poppy, lousewort, *Erythronium* and *Erythrina crista-galli*; the characteristic jester's cap; a shape of pasta (*creste di galli*); and so on.

Spelling Variations

- cockscomb
- cock's-comb
- cock's comb
- coxcomb

Lore (Anatomy)

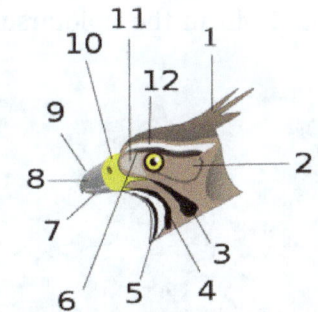

Arrow "11" indicates the lore of this eagle

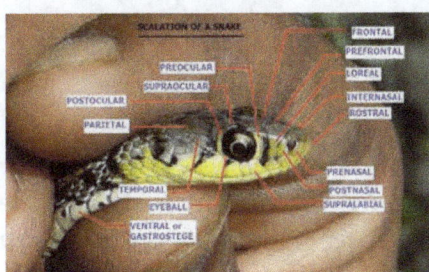

Scales on a snake's head

The lore is the region between the eyes and nostrils of birds, reptiles, and amphibians.

Ornithology

In ornithology, the lore is the region between the eye and bill on the side of a bird's head. This region is sometimes featherless, and the skin may be tinted, as in many species of the cormorant family. This area, which is directly in front of the eye, features a "loral stripe" in many bird species including the red-capped plover.

Herpetology

In amphibians and reptiles, lore pertains to the regions immediately adjacent to the eyes and between the eyes and nostrils. These are analogous to the lore on birds which corresponds to the region between the eye and the beak.

In snakes and reptiles, a loreal scale also refers to the scales which lie between the eye and the nostril. In crotaline snakes (pit vipers), loreal pits are present on either side of the head.

Bird Feet and Legs

The anatomy of bird legs and feet is very diverse, revealing many accommodations to perform a wide variety of functions.

Most birds are digitigrade animals, which means that they walk on their toes, not the entire foot.

Some of the lower bones of the foot (the distals and most of the metatarsal) are fused to form the tarsometatarsus – a third segment of the leg, specific to birds. The upper bones of the foot (proximals) in turn are fused with the tibia to form the tibiotarsus, as over time the centralia disappeared. The fibula is also reduced.

African jacana. Extremely long toes and claws help distribute the jacana's weight over a wide area to allow it to walk on floating leaves.

The legs are attached to a very strong assembly consisting of the pelvic girdle extensively fused with the uniform spinal bone (also specific to birds) called the *synsacrum*, built from some of the fused bones.

Skeleton

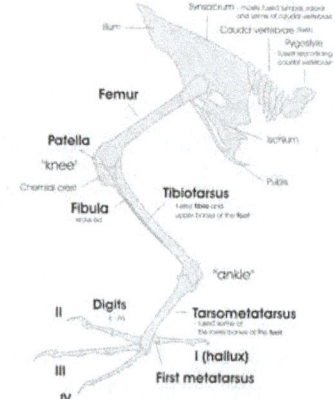

Bird left leg and pelvic girdle skeleton

Hindlimbs

Birds are generally digitigrade animals (toe-walkers) as reflected in the structure of their leg skeleton. They use only their hindlimbs to walk, which is called bipedalism; their forelimbs have evolved to become wings. As mentioned earlier, most bones of the avian foot (except toes) are fused together or with other bones, having over time changed function.

Tarsometatarsus

Some lower bones of the foot are fused to form the tarsometatarsus – a third segment of the leg specific for birds. It consists of merged distals and metatarsals II, III and IV. Metatarsus I remains separated as a base of the first toe. The tarsometatarsus is the extended foot area giving the leg an extra lever length.

Tibiotarsus

The upper bones of the foot (proximals) are fused with the tibia to form the tibiotarsus, while the centralia disappeared. The anterior (frontal) side of the dorsal end of the tibiotarsus (at the knee) contains a protruding enlargement called the cnemial crest.

Patella

At the knee above the cnemial crest is the patella (kneecap). Some species do not have patellas; sometimes it is only a small extension of the cnemial crest. In grebes there exist both a normal patella and an extension of the cnemial crest.

Fibula

The fibula is reduced and adheres extensively to the tibia, usually reaching only two-thirds of its length. Only penguins have full-length fibulae.

Femur

The femur is normal.

Knee and Ankle - Confusions

The bird knee joint between the femur and tibia (or rather tibiotarsus) points forwards, but is hidden within the feathers. The backward-pointing "heel" (ankle) we can see is a joint between the tibiotarsus and tarsometatarsus. The joint inside the tarsus occurs also in some reptiles. It is worth noting here that the name "thick knee" of the birds in the family Burhinidae is incorrect, because the heel of these birds is large. So ornithologists confused knee and ankle here.

Toes and Unfused Metatarsals

Most birds have four toes (digits), typically three facing forward and one pointing backward. In a typical perching bird they consist respectively of 3, 4, 5 and 2 phalanges. Some birds have only the three forward-facing toes (tridactyl, for example the sanderling) and the ostrich has only two toes (didactyl). The first digit, called the hallux, is homologous to the human big toe and usually projects to the rear.

On the extreme phalanx of each toe are located the claws. They consist of a horny keratinous podotheca or sheath, so are not part of the skeleton.

The bird foot also contains one or two metatarsals not fused in the tarsometatarsus.

Pelvic Girdle and Synsacrum

The legs are attached to a very strong (but lightweight) assembly consisting of the pelvic girdle extensively fused with the uniform spinal bone called the synsacrum, which is specific to birds. The synsacrum is built from fused lumbar, sacral, some of first caudal and sometimes the last one or two thoracic vertebrae, depending on species (birds have altogether between 10 and 22 vertebrae). Except for ostriches and rheas, pubic bones in the pelvic girdle do not connect each other, making egg-laying easier.

Rigidity and Reduction of Mass

Fusions of individual bones into rigid strong structures are very characteristic for the bird skeleton.

Most major bird bones are also extensively pneumatized. They contain many air pockets connected to the pulmonary air sacs of the respiratory system. Their spongy interior makes them very strong relatively to their mass. The number of pneumatic bones however depends on species. It is worth noting that the pneumaticity is slight or absent in diving birds. For example, in the long-tailed duck, the bones of the leg (and wing) are not pneumatic in contrast with some of the other bones, while loons and puffins have even more massive skeletons without any aired bones. The flightless ostrich and emu have pneumatic femurs and so far this is the only known pneumatic bone in these birds except cervical vertebrae of the ostrich.

Fusions (rigidity) and pneumatic bones (reduction of mass) are considered as some of the many adaptations of birds for flight.

Plantigrade Locomotion in Birds

Most birds except loons and grebes are digitigrade, not plantigrade. Also chicks in the nest can use the entire foot (toes and tarsometatarsus) with the heel on the ground.

Loons tend to walk this way because their legs and pelvis are deeply specialized for swimming. They have a narrow pelvis, which moves the attachment point of the femur to the rear, and their tibiotarsus is much longer than the femur. This shifts the feet (toes) behind the center of mass of the loon body. They walk usually by pushing themselves on their breasts; larger loons cannot take off from land. This position however is highly suitable for swimming, because their feet are located at the rear like the propeller on a motorboat.

Grebes and many other waterfowl have shorter femur and more or less narrow pelvis too, which gives the impression that their legs are attached to the rear like in loons.

Functions of Bird Legs and Feet

African grey parrot grips the perch with zygodactyl feet

Palmate feet – Chilean flamingo

Totipalmate feet – blue-footed booby

Western grebe presenting a lobate foot

Lobate feet – a chick of the Eurasian coot

The snowshoelike foot of the willow ptarmigan is an adaptation to walking on snow

Because avian forelimbs are converted to wings, many of their functions are performed by the bill and hindlimbs (feet and legs). It has been proposed that the hindlimbs are also very important in flight as accelerators when taking-off. Some leg and foot functions, including conventional ones and those specific to birds are:

- locomotion
  - o walking, running (ostrich, grouse, wild turkey), hopping, climbing (woodpeckers, nut-hatches, treecreepers)
  - o swimming and steering underwater (ducks, grebes, loons)
- perching on a branch or clinging
- carrying (like osprey holding fish)
- flight - related
  - o serving probably as the primary take-off accelerator. In the common vampire bat in contrast the required force is generated by the forelimb (the wing).
  - o absorbing the shock of landing on a perch and on the water, becoming "water skis"
- feeding - related
  - o catching and killing prey in raptors (hawk, owl)
  - o holding (parrots - used like hands) and pulling apart food (along with the bill)
  - o scratching the ground in search of food
    - □ "double scratch" - hopping forward and then backward using both feet to scratch (often towhee, sparrow, junco)
    - □ one-footed scratch (grouse, quail, wild turkey, domestic chicken)

- reproduction - related

  o cradling and turning eggs during incubation. Birds lacking a brood patch incubate the eggs with their feet - grasping one or even two of them (gannets, boobies) or keeping them on the top surfaces of their feet (penguins under a pouch of belly skin, murres).

  o courtship (sage grouse), aerial courtship (bald eagle)

  o building nests (in addition to the bill)

- preening and cleaning. Sometimes birds use a special claw, for example barn owls have a so-called "feather comb", some herons and nightjars use the claw for cleaning the head.

- heat loss regulation (herons, gulls, giant petrel, storks, New World vultures, ducks, geese)

Toe Arrangements

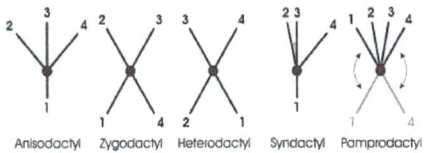

Bird right foot - toe arrangement

Typical toe arrangements in birds are:

- anisodactyl - three toes in front (2, 3, 4), one in back (1)

  Nearly all songbirds and most other perching birds.

- zygodactyl - two toes in front (2, 3) and two in back (1, 4) - the outermost front toe (4) is reversed

  Many perching birds - most woodpeckers and their allies, osprey, owls, cuckoos, most parrots, mousebirds, some swifts, cuckoo roller.

  Woodpeckers when climbing can rotate outer rear digit (4) to the side in an ectropodactyl arrangement. Black-backed woodpecker, Eurasian three-toed woodpecker and American three-toed woodpecker have three toes - the inner rear (1) is missing, outer rear (4) points always backward (never rotates).

  Owls, osprey and turacos can rotate outer toe (4) back and forth.

  The zygodactyl arrangement is a case of convergence, because it evolved in birds in different ways nine times.

- heterodactyl - two toes in front (3, 4) and two in back (2, 1) - the inner front toe (2) is reversed.

  Only present in trogons.

- syndactyl - three toes in front (2, 3, 4), one in back (1), two - outer and middle (2, 3) are joined for much of their length.

  Common in Coraciiformes, including kingfishers and hornbills.

- pamprodactyl - two inner toes in front (2, 3), two outer (1, 4) can rotate freely forward and backward.

  Mousebirds, some swifts.

  Some swifts move all four digits forward to use them as hooks to hang.

Most common is the anisodactyl foot, second among perching birds is the zygodactyl arrangement.

Webbing and Lobation

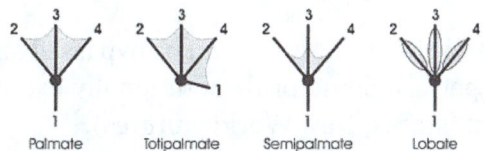

Bird feet - webbing and lobation (right foot)

Palmations and lobes enable swimming or help walking on loose ground like over mud. The webbed or palmated feet of birds can be categorized into the following types:

- palmate - only the anterior digits (2-4) joined by the webbing.

  Ducks, geese and swans, gulls and terns, and other aquatic birds (auks, flamingos, fulmars, jaegers, loons, petrels, shearwaters, skimmers).

  Diving ducks additionally have lobed hind toe (1), gulls, terns and allies have reduced hind toe.

- totipalmate - all four digits (1-4) joined by the webbing.

  Gannets and boobies, pelicans, cormorants, anhingas, frigatebirds.

  Some gannets have brightly colored feet used in display.

- semipalmate - a small web between the anterior digits (2-4).

  Some plovers (Eurasian dotterel) and sandpipers (semipalmated sandpiper, stilt sandpiper, upland sandpiper, greater yellowlegs, willet), avocet, herons (only two toes), all grouse, some domestic breeds of chicken.

  Plovers and lapwings have a vestigial hind toe (1), sandpipers and their allies a reduced and raised hind toe almost not touching the ground. Sanderling is the only sandpipper having 3 toes (tridactyl foot).

- lobate - the anterior digits (2-4) edged with lobes of skin.

  Lobes expand or contract when bird swims.

  Grebes, coots, phalaropes, finfoots, some palmate-footed ducks on the hallux (1).

  Grebes have more webbing between the toes in comparison to coots and phalaropes.

Most common is the palmate foot.

Heat Loss Regulation

Some of birds like gulls, herons, ducks or geese can regulate heat loss by their feet.

The arteries (carrying blood from the heart) and veins (carrying blood toward the heart) inter-twine in the legs so heat can be transferred back from arteries to veins before reaching feet. Such a mechanism is called countercurrent exchange. Gulls can also open a shunt between these vessels, turning back the blood stream above the foot, and constrict the vessels in the foot. This reduces heat loss by more than 90 percent. In gulls the temperature of the base of the leg is 32°C (89°F), of the foot may be close to 0°C (32°F).

However, for cooling, this heat-exchange network can be bypassed and blood flow through the foot significantly increased (giant petrel). Some birds additionally excrete onto their feet, increasing heat loss thanks to evaporation (storks, New World vultures).

Beak

Comparison of bird beaks, displaying different shapes adapted to different feeding methods. Not to scale.

The beak, bill, or rostrum is an external anatomical structure of birds that is used for eating and for grooming, manipulating objects, killing prey, fighting, probing for food, courtship and feeding young. The terms *beak* and *rostrum* are also used to refer to a similar mouth part in some dicynodonts, Ornithischians, cephalopods, cetaceans, billfishes, pufferfishes, turtles, Anuran tadpoles and sirens.

Although beaks vary significantly in size, shape, color and texture, they share a similar underlying structure. Two bony projections—the upper and lower mandibles—are covered with a thin keratinized layer of epidermis known as the rhamphotheca. In most species, two holes known as nares lead to the respiratory system.

Etymology

Although the word *beak* was, in the past, generally restricted to the sharpened bills of birds of prey, in modern ornithology, the terms *beak* and *bill* are generally considered to be synonymous. The word, which dates from the 13th century, comes from the Middle English *bec*, which itself comes from the Latin *beccus*.

Anatomy

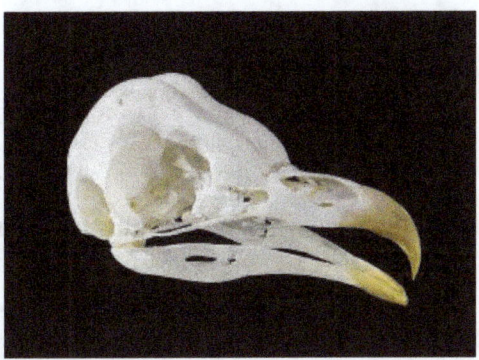

The bony core of the beak is a lightweight framework, like that seen on this barn owl's skull.

Although beaks vary significantly in size and shape from species to species, their underlying structures have a similar pattern. All beaks are composed of two jaws, generally known as the upper mandible (or maxilla) and lower mandible (or mandible). The upper, and in some cases the lower, mandibles are strengthened internally by a complex three-dimensional network of bony spicules (or trabeculae) seated in soft connective tissue and surrounded by the hard outer layers of the beak. The avian jaw apparatus is made up of two units: one four-bar linkage mechanism and one five-bar linkage mechanism.

Mandibles

A gull's upper mandible can flex upwards because it is supported by small bones which can move slightly backwards and forwards.

The upper mandible is supported by a three-pronged bone called the intermaxillary. The upper prong of this bone is embedded into the forehead, while the two lower prongs attach to the sides of the skull. At the base of the upper mandible a thin sheet of nasal bones is attached to the skull at the nasofrontal hinge, which gives mobility to the upper mandible, allowing it to move upwards and downwards.

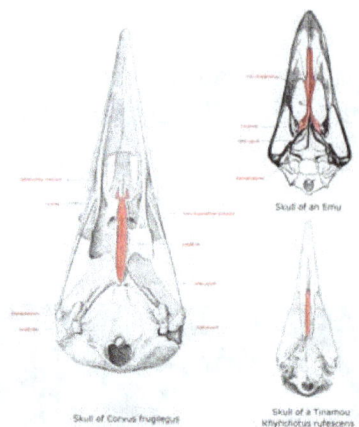

Position of vomer (shaded red) in neognathae (left) and paleognathae (right)

The base of the upper mandible, or the roof when seen from the mouth, is the palate, the structure of which differs greatly in the ratites. Here, the vomer is large and connects with premaxillae and maxillopalatine bones in a condition termed as a "paleognathous palate". All other extant birds have a narrow forked vomer that does not connect with other bones and is then termed as neognathous. The shape of these bones varies across the bird families.

The lower mandible is supported by a bone known as the inferior maxillary bone—a compound bone composed of two distinct ossified pieces. These ossified plates (or rami), which can be U-shaped or V-shaped, join distally (the exact location of the joint depends on the species) but are separated proximally, attaching on either side of the head to the quadrate bone. The jaw muscles, which allow the bird to close its beak, attach to the proximal end of the lower mandible and to the bird's skull. The muscles that depress the lower mandible are usually weak, except in a few birds such as the starlings and the extinct Huia, which have well-developed digastric muscles that aid in foraging by prying or gaping actions. In most birds, these muscles are relatively small as compared to the jaw muscles of similarly sized mammals.

Rhamphotheca

The outer surface of the beak consists of a thin horny sheath of keratin called the rhamphotheca, which can be subdivided into the rhinotheca of the upper mandible and the gnathotheca of the lower mandible. This covering arises from the Malpighian layer of the bird's epidermis, growing from plates at the base of each mandible. There is a vascular layer between the rhamphotheca and the deeper layers of the dermis, which is attached directly to the periosteum of the bones of the beak. The rhamphotheca grows continuously in most birds, and in some species, the color varies seasonally. In some alcids, such as the puffins, parts of the rhamphotheca are shed each year after the breeding season, while some pelicans shed a part of the bill called a "bill horn" that develops in the breeding season.

While most extant birds have a single seamless rhamphotheca, species in a few families, including the albatrosses and the emu, have compound rhamphothecae that consist of several pieces separated and defined by softer keratinous grooves. Studies have shown that this was the primitive ancestral state of the rhamphotheca, and that the modern simple rhamphotheca resulted from the gradual loss of the defining grooves through evolution.

Tomia

The sawtooth serrations on a common merganser's bill help it to hold tight to its fish prey.

The tomia (singular *tomium*) are the cutting edges of the two mandibles. In most birds, these range from rounded to slightly sharp, but some species have evolved structural modifications that allow them to handle their typical food sources better. Granivorous (seed-eating) birds, for example, have ridges in their tomia, which help the bird to slice through a seed's outer hull. Most falcons have a sharp projection along the upper mandible, with a corresponding notch on the lower mandible. They use this "tooth" to sever their prey's vertebrae fatally or to rip insects apart. Some kites, principally those that prey on insects or lizards, also have one or more of these sharp projections, as do the shrikes. Some fish-eating species, e.g., the mergansers, have sawtooth serrations along their tomia, which help them to keep hold of their slippery, wriggling prey.

Birds in roughly 30 families have tomia lined with tight bunches of very short bristles along their entire length. Most of these species are either insectivores (preferring hard-shelled prey) or snail eaters, and the brush-like projections may help to increase the coefficient of friction between the mandibles, thereby improving the bird's ability to hold hard prey items. Serrations on hummingbird bills, found in 23% of all hummingbird genera, may perform a similar function, allowing the birds to effectively hold insect prey. They may also allow shorter-billed hummingbirds to function as nectar thieves, as they can more effectively hold and cut through long or waxy flower corollas. In some cases, the color of a bird's tomia can help to distinguish between similar species. The snow goose, for example, has a reddish-pink bill with black tomia, while the whole beak of the similar Ross's goose is pinkish-red, without darker tomia.

Culmen

A thrush's culmen is measured in a straight line from the tip of the beak to a set point — here, where the feathering starts on the bird's forehead.

The culmen is the dorsal ridge of the upper mandible. Likened by ornithologist Elliott Coues to the ridge line of a roof, it is the "highest middle lengthwise line of the bill" and runs from the point where the upper mandible emerges from the forehead's feathers to its tip. The bill's length along the culmen is one of the regular measurements made during bird banding (ringing) and is particularly useful in feeding studies. There are several standard measurements that can be made—from the beak's tip to the point where feathering starts on the forehead, from the tip to the anterior edge of the nostrils, from the tip to the base of the skull, or from the tip to the cere (for raptors and owls)—and scientists from various parts of the world generally favor one method over another. In all cases, these are chord measurements (measured in a straight line from point to point, ignoring any curve in the culmen) taken with calipers.

The shape or color of the culmen can also help with the identification of birds in the field. For example, the culmen of the parrot crossbill is strongly decurved, while that of the very similar red crossbill is more moderately curved. The culmen of a juvenile common loon is all dark, while that of the very similarly plumaged juvenile yellow-billed loon is pale towards the tip.

Gonys

The gonys is the ventral ridge of the lower mandible, created by the junction of the bone's two rami, or lateral plates. The proximal end of that junction—where the two plates separate—is known as the gonydeal angle or gonydeal expansion. In some gull species, the plates expand slightly at that point, creating a noticeable bulge; the size and shape of the gonydeal angle can be useful in identifying between otherwise similar species. Adults of many species of large gulls have a reddish or orangish gonydeal spot near the gonydeal expansion. This spot triggers begging behavior in gull chicks. The chick pecks at the spot on its parent's bill, which in turn stimulates the parent to regurgitate food.

Commissure

Depending on its usage, *commissure* may refer to the junction of the upper and lower mandibles, or alternately, to the full-length apposition of the closed mandibles, from the corners of the mouth to the tip of the beak.

Gape

A young starling with a bright yellow gape

In bird anatomy, the gape is the interior of the open mouth of a bird, and the gape flange is the region where the two mandibles join together at the base of the beak. The width of the gape can be a factor in the choice of food.

The gape flange on this juvenile house sparrow is the yellowish region at the base of the beak

Gapes of juvenile altricial birds are often brightly coloured, sometimes with contrasting spots or other patterns, and these are believed to be an indication of their health, fitness and competitive ability. Based on this, the parents decide how to distribute food among the chicks in the nest. Some species, especially in the families Viduidae and Estrildidae, have bright spots on the gape known as gape tubercles or gape papillae. These nodular spots are conspicuous even in low light. A study examining the nestling gapes of eight passerine species found that the gapes were conspicuous in the ultraviolet spectrum (visible to birds but not to humans). Parents may, however, not rely solely on the gape coloration, and other factors influencing their decision remain unknown.

Red gape color has been shown in several experiments to induce feeding. An experiment in manipulating brood size and immune system with barn swallow nestlings showed the vividness of the gape was positively correlated with T-cell–mediated immunocompetence, and that larger brood size and injection with an antigen led to a less vivid gape. Conversely, the red gape of the common cuckoo (*Cuculus canorus*) did not induce extra feeding in host parents. Some brood parasites, such as the Hodgson's hawk-cuckoo (*C. fugax*), have colored patches on the wing that mimic the gape color of the parasitized species.

When born, the chick's gape flanges are fleshy. As it grows into a fledgling, the gape flanges remain somewhat swollen and can thus be used to recognize that a particular bird is young. By the time it reaches adulthood, the gape flanges will no longer be visible.

Nares

Falcons have a small tubercule within each nare

Most species of birds have external nares (nostrils) located somewhere on their beak. The nares are two holes—circular, oval or slit-like in shape—which lead to the nasal cavities within the bird's skull, and thus to the rest of the respiratory system. In most bird species, the nares are located in the basal third of the upper mandible. Kiwis are a notable exception; their nares are located at the tip of their bills. A handful of species have no external nares. Cormorants and darters have primitive external nares as nestlings, but these close soon after the birds fledge; adults of these species (and gannets and boobies of all ages, which also lack external nostrils) breathe through their mouths. There is typically a septum made of bone or cartilage that separates the two nares, but in some families (including gulls, cranes and New World vultures), the septum is missing. While the nares are uncovered in most species, they are covered with feathers in a few groups of birds, including grouse and ptarmigans, crows, and some woodpeckers. The feathers over a ptarmigan's nostrils help to warm the air it inhales, while those over a woodpecker's nares help to keep wood particles from clogging its nasal passages.

Species in the bird order Procellariformes have nostrils enclosed in double tubes which sit atop or along the sides of the upper mandible. These species, which include the albatrosses, petrels, diving petrels, storm petrels, fulmars and shearwaters, are widely known as "tubenoses". A number of species, including the falcons, have a small bony tubercule which projects from their nares. The function of this tubercule is unknown. Some scientists suggest it may act as a baffle, slowing down or diffusing airflow into the nares (and thus allowing the bird to continue breathing without damaging its respiratory system) during high-speed dives, but this theory has not been proved experimentally. Not all species that fly at high speeds have such tubercules, while some species which fly at low speeds do.

Operculum

The rock dove's operculum is a mass at the base of the bill

The nares of some birds are covered by an operculum (plural *opercula*), a membraneous, horny or cartilaginous flap. In diving birds, the operculum keeps water out of the nasal cavity; when the birds dive, the impact force of the water closes the operculum. Some species which feed on flowers have opercula to help to keep pollen from clogging their nasal passages, while the opercula of the two species of *Attagis* seedsnipe help to keep dust out. The nares of nestling tawny frogmouths are covered with large dome-shaped opercula, which help to reduce the rapid evaporation of water vapor, and may also help to increase condensation within the nostrils themselves—both critical functions, since the nestlings get fluids only from the food their parents bring them. These opercula shrink as the birds age, disappearing completely by the time they reach adulthood. In pigeons, the operculum has evolved into a soft swollen mass that sits at the base of the bill, above the nares; though it is sometimes referred to as the *cere*, this is a different structure. Tapaculos are the only birds able to move their opercula.

Rosette

Some species, such as the puffin, have a fleshy rosette, sometimes called a "gape rosette", at the corners of the beak. In the puffin, this is grown as part of its display plumage.

Cere

Birds from a handful of families—including raptors, owls, skuas, parrots, turkeys and curassows— have a waxy structure called a cere (from the Latin *cera*, which means "wax") which covers the base of their bill. This structure typically contains the nares, except in the owls, where the nares are distal to the cere. Although it is sometimes feathered in parrots, the cere is typically bare and often brightly colored. In raptors, the cere is a sexual signal which indicates the "quality" of a bird; the orangeness of a Montague's harrier's cere, for example, correlates to its body mass and physical condition. The cere color of young Eurasian scops-owls has an ultraviolet (UV) component, with a UV peak that correlates to the bird's mass. A chick with a lower body mass has a UV peak at a higher wavelength than a chick with a higher body mass does. Studies have shown that parent owls preferentially feed chicks with ceres that show higher wavelength UV peaks, that is, lighter-weight chicks.

The color or appearance of the cere can be used to distinguish between males and females in some species. For example, the male great curassow has a yellow cere, which the female (and young males) lack. The male budgerigar's cere is blue, while the female's is pinkish or brown.

Nail

The nail is the black tip of this mute swan's beak

All birds of the family Anatidae (ducks, geese, and swans) have a nail, a plate of hard horny tissue at the tip of the beak. This shield-shaped structure, which sometimes spans the entire width of the beak, is often bent at the tip to form a hook. It serves different purposes depending on the bird's primary food source. Most species use their nails to dig seeds out of mud or vegetation, while diving ducks use theirs to pry molluscs from rocks. There is evidence that the nail may help a bird to grasp things; species which use strong grasping motions to secure their food (such as when catching and holding onto a large squirming frog) have very wide nails. Certain types of mechano-receptors, nerve cells that are sensitive to pressure, vibration or touch, are located under the nail.

The shape or color of the nail can sometimes be used to help distinguish between similar-looking species or between various ages of waterfowl. For example, the greater scaup has a wider black nail than does the very similar lesser scaup. Juvenile "grey geese" have dark nails, while most adults have pale nails. The nail gave the wildfowl family one of its former names: "Unguirostres" comes from the Latin *ungus*, meaning "nail" and *rostrum*, meaning "beak".

Rictal Bristles

Rictal bristles are stiff hair-like feathers that arise around the base of the beak. They are common among insectivorous birds, but are also found in some non-insectivorous species. Their function is uncertain, although several possibilities have been proposed. They may function as a "net", helping in the capture of flying prey, although to date, there has been no empirical evidence to support this idea. There is some experimental evidence to suggest that they may prevent particles from striking the eyes if, for example, a prey item is missed or broken apart on contact. They may also help to protect the eyes from particles encountered in flight, or from casual contact from vegetation. There is also evidence that the rictal bristles of some species may function tactilely, in a manner similar to that of mammalian whiskers (vibrissae). Studies have shown that Herbst corpuscles, mechanoreceptors sensitive to pressure and vibration, are found in association with rictal bristles. They may help with prey detection, with navigation in darkened nest cavities, with the gathering of information during flight or with prey handling.

Egg Tooth

This Arctic tern chick still has its egg tooth, the small white projection near the tip of its upper mandible.

Full-term chicks of most bird species have a small sharp, calcified projection on their beak, which they use to chip their way out of their egg. Commonly known as an egg tooth, this white spike is generally near the tip of the upper mandible, though some species have one near the tip of their lower mandible instead, and a few species have one on each mandible. Despite its name, the projection is not an actual tooth, as the similarly-named projections of some reptiles are; instead, it is part of the integumentary system, as are claws and scales. The hatching chick first uses its egg tooth to break the membrane around an air chamber at the wide end of the egg. Then it pecks at the eggshell while turning slowly within the egg, eventually (over a period of hours or days) creating a series of small circular fractures in the shell. Once it has breached the egg's surface, the chick continues to chip at it until it has made a large hole. The weakened egg eventually shatters under the pressure of the bird's movements. The egg tooth is so critical to a successful escape from the egg that chicks of most species will perish unhatched if they fail to develop one. However, there are a few species which do not have egg teeth. Megapode chicks have an egg tooth while still in the egg but lose it before hatching, while kiwi chicks never develop one; chicks of both families escape their eggs by kicking their way out. Most chicks lose their egg teeth within a few days of hatching, though petrels keep theirs for nearly three weeks and marbled murrelets have theirs for up to a month. Generally, the egg tooth drops off, though in songbirds it is reabsorbed.

Color

The color of a bird's beak results from concentrations of pigments—primarily melanins and carotenoids—in the epidermal layers, including the rhamphotheca. Eumelanin, which is found in the bare parts of many bird species, is responsible for all shades of gray and black; the denser the deposits of pigment found in the epidermis, the darker the resulting color. Phaeomelanin produces "earth tones" ranging from gold and rufous to various shades of brown. Although it is thought to occur in combination with eumelanin in beaks which are buff, tan, or horn-colored, researchers have yet to isolate phaeomelanin from any beak structure. More than a dozen types of carotenoids are responsible for the coloration of most red, orange, and yellow beaks. The hue of the color is determined by the precise mix of red and yellow pigments, while the saturation is determined by the density of the deposited pigments. For example, bright red is created by dense deposits of mostly red pigments, while dull yellow is created by diffuse deposits of mostly yellow pigments. Bright orange is created by dense deposits of both red and yellow pigments, in roughly equal concentrations. Beak coloration helps to make displays using those beaks more obvious.

Birds are capable of seeing colors in the ultraviolet range, and some species are known to have ultraviolet peaks of reflectance (indicating the presence of ultraviolet color) on their beaks. The presence and intensity of these peaks may indicate a bird's fitness, sexual maturity or pair bond status. King and emperor penguins, for example, show spots of ultraviolet reflectance only as adults. These spots are brighter on paired birds than on courting birds. The position of such spots on the beak may be important in allowing birds to identify conspecifics. For instance, the very similarly-plumaged king and emperor penguins have UV-reflective spots in different positions on their beaks.

In general, beak color depends on a combination of the bird's hormonal state and diet. Colors are typically brightest as the breeding season approaches, and palest after breeding.

Dimorphism

The size and shape of the beak can vary across species as well as between them; in some species, the size and proportions of the beak vary between males and females. This allows the sexes to utilize different ecological niches, thereby reducing intraspecific competition. For example, females of nearly all shorebirds have longer bills than males of the same species, and female American avocets have beaks which are slightly more upturned than those of males. Males of the larger gull species have bigger, stouter beaks than those of females of the same species, and immatures can have smaller, more slender beaks than those of adults. Many hornbills show sexual dimorphism in the size and shape of both beaks and casques, and the female huia's slim, decurved bill was nearly twice as long as the male's straight, thicker one.

The beaks of the now-extinct Huia (female upper, male lower) show marked sexual dimorphism

Color can also differ between sexes or ages within a species. Typically, such a color difference is due to the presence of androgens. For example, in house sparrows, melanins are produced only in the presence of testosterone; castrated male house sparrows—like female house sparrows—have brown beaks. Castration also prevents the normal seasonal color change in the beaks of male black-headed gulls and indigo buntings.

Functions

Birds may bite or stab with their beaks to defend themselves. Some species use their beaks in displays of various sorts. As part of his courtship, for example, the male garganey touches his beak to the blue speculum feathers on his wings in a fake preening display, and the male Mandarin duck does the same with his orange sail feathers. A number of species use a gaping, open beak in their fear and/or threat displays. Some augment the display by hissing or breathing heavily, while others clap their beaks.

Communication

A number of species, including storks, some owls, frogmouths and the noisy miner, use bill clapping as a form of communication.

Heat Exchange

Studies have shown that some birds use their beaks to rid themselves of excess heat. The toco toucan, which has the largest beak relative to the size of its body of any bird species, is capable of modifying the blood flow to its beak. This process allows the beak to work as a "transient thermal radiator", reportedly rivaling an elephant's ears in its ability to radiate body heat. Measurements of the bill sizes of several species of American sparrows found in salt marshes along the North American coastlines show a strong correlation with summer temperatures recorded in the locations where the sparrows breed; latitude alone showed a much weaker correlation. By dumping excess heat through their bills, the sparrows are able to avoid the water loss which would be required by evaporative cooling—an important benefit in a windy habitat where freshwater is scarce. Several ratites, including the common ostrich, the emu and the southern cassowary, use various bare parts of their bodies (including their beaks) to dissipate as much as 40% of their metabolic heat production. Alternately, studies have shown that birds from colder climates (higher altitudes or latitudes and lower environmental temperatures) have smaller beaks, lessening heat loss from that structure.

Billing

When billing, northern gannets raise their beaks high and clatter them against each other.

During courtship, mated pairs of many bird species touch or clasp each other's bills. Termed billing (also nebbing in British English), this behavior appears to strengthen pair bonding. The amount of contact involved varies among species. Some gently touch only a part of their partner's beak while others clash their beaks vigorously together.

Gannets raise their bills high and repeatedly clatter them, the male puffin nibbles at the female's beak, the male waxwing puts his bill in the female's mouth and ravens hold each other's beaks in a prolonged "kiss". Billing can also be used as a gesture of appeasement or subordination. Subordinate gray jay routinely bill more dominant birds, lowering their body and quivering their wings in the manner of a young bird food begging as they do so. A number of parasites, including rhinonyssids and *Trichomonas gallinae* are known to be transferred between birds during episodes of billing.

Usage of the term has spread beyond avian behavior; "billing and cooing" in reference to human courtship (particularly kissing) has been in use since Shakespeare's time, and derives from the courtship of doves.

Beak Trimming

Because the beak is a sensitive organ with many sensory receptors, beak trimming (sometimes referred to as 'debeaking') is "acutely painful" to the birds it is performed on. It is nonetheless routinely done to intensively farmed poultry flocks, particularly laying and broiler breeder flocks, because it helps reduce the damage the flocks inflict on themselves due to a number of stress-induced behaviors, including cannibalism, vent pecking and feather pecking. A cauterizing blade or infrared beam is used to cut off about half of the upper beak and about a third of the lower beak.

Pain and sensitivity can persist for weeks or months after the procedure, and neuromas can form along the cut edges. Food intake typically decreases for some period after the beak is trimmed. However, studies show that trimmed poultry's adrenal glands weigh less, and their plasma corticosterone levels are lower than those found in untrimmed poultry, indicating that they are less stressed overall.

A similar but separate practice, usually performed by an avian veterinarian or an experienced birdkeeper, involves clipping, filing or sanding the beaks of captive birds for health purposes – in order to correct or temporarily alleviate overgrowths or deformities and better allow the bird to go about its normal feeding and preening activities. Amongst raptor keepers, this practice is commonly known as "coping".

Bill Tip Organ

The bill tip organ is a region found near the tip of the bill in several types of birds that forage particularly by probing. The region has a high density of nerve endings known as the corpuscles of Herbst. This consists of pits in the bill surface which in the living bird is occupied by cells that sense pressure changes. The assumption is that this allows the bird to perform 'remote touch', which means that it can detect movements of animals which the bird does not directly touch. Bird species known to have a 'bill-tip organ' includes members of ibisis, shorebirds of the family Scolopacidae, and kiwis.

There is a suggestion that across these species, the bill tip organ is more well developed among species foraging in wet habitats (water column or soft mud) than in species using a more terrestrial foraging. However, it has been described in terrestrial birds too, including parrots, who are known for their dextrous extractive foraging techniques. Unlike probing foragers, the tactile pits in parrots are embedded in the hard keratin (or rhamphotheca) of the bill, rather than the bone, and along the inner edges of the curved bill, rather than being on the outside of the bill.

Bird Vision

With forward-facing eyes, the bald eagle has a wide field of binocular vision.

Vision is the most important sense for birds, since good eyesight is essential for safe flight, and this group has a number of adaptations which give visual acuity superior to that of other vertebrate groups; a pigeon has been described as "two eyes with wings". The avian eye resembles that of a reptile, with ciliary muscles that can change the shape of the lens rapidly and to a greater extent than in the mammals. Birds have the largest eyes relative to their size in the animal kingdom, and movement is consequently limited within the eye's bony socket. In addition to the two eyelids usually found in vertebrates, it is protected by a third transparent movable membrane. The eye's internal anatomy is similar to that of other vertebrates, but has a structure, the pecten oculi, unique to birds.

Some bird groups have specific modifications to their visual system linked to their way of life. Birds of prey have a very high density of receptors and other adaptations that maximise visual acuity. The placement of their eyes gives them good binocular vision enabling accurate judgement of distances. Nocturnal species have tubular eyes, low numbers of colour detectors, but a high density of rod cells which function well in poor light. Terns, gulls and albatrosses are amongst the seabirds which have red or yellow oil droplets in the colour receptors to improve distance vision especially in hazy conditions.

Extraocular Anatomy

The eye of a bird most closely resembles that of the reptiles. Unlike the mammalian eye, it is not spherical, and the flatter shape enables more of its visual field to be in focus. A circle of bony plates, the sclerotic ring, surrounds the eye and holds it rigid, but an improvement over the reptilian eye, also found in mammals, is that the lens is pushed further forward, increasing the size of the image on the retina.

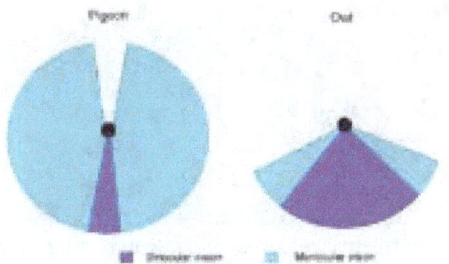

Visual fields for a pigeon and an owl

Most birds cannot move their eyes, although there are exceptions, such as the great cormorant. Birds with eyes on the sides of their heads have a wide visual field, useful for detecting predators, while those with eyes on the front of their heads, such as owls, have binocular vision and can estimate distances when hunting. The American woodcock probably has the largest visual field of any bird, 360° in the horizontal plane, and 180° in the vertical plane.

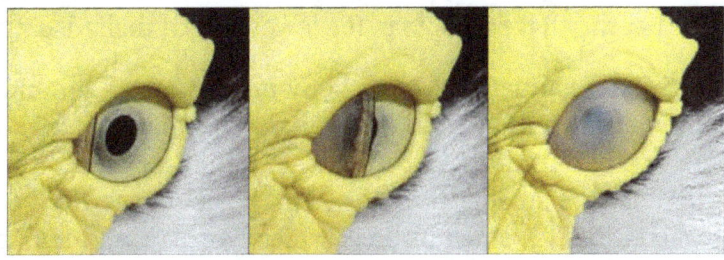

The nictitating membrane of a masked lapwing

The eyelids of a bird are not used in blinking. Instead the eye is lubricated by the nictitating membrane, a third concealed eyelid that sweeps horizontally across the eye like a windscreen wiper. The nictitating membrane also covers the eye and acts as a contact lens in many aquatic birds when they are under water. When sleeping, the lower eyelid rises to cover the eye in most birds, with the exception of the horned owls where the upper eyelid is mobile.

The eye is also cleaned by tear secretions from the lachrymal gland and protected by an oily substance from the Harderian glands which coats the cornea and prevents dryness. The eye of a bird is larger compared to the size of the animal than for any other group of animals, although much of it is concealed in its skull. The ostrich has the largest eye of any land vertebrate, with an axial length of 50 mm (2 in), twice that of the human eye.

Bird eye size is broadly related to body mass. A study of five orders (parrots, pigeons, petrels, raptors and owls) showed that eye mass is proportional to body mass, but as expected from their habits and visual ecology, raptors and owls have relatively large eyes for their body mass.

Behavioural studies show that many avian species focus on distant objects preferentially with their lateral and monocular field of vision, and birds will orientate themselves sideways to maximise visual resolution. For a pigeon, resolution is twice as good with sideways monocular vision than forward binocular vision, whereas for humans the converse is true.

The European robin has relatively large eyes, and starts to sing early in the morning.

The performance of the eye in low light levels depends on the distance between the lens and the retina, and small birds are effectively forced to be diurnal because their eyes are not large enough to give adequate night vision. Although many species migrate at night, they often collide with even brightly lit objects like lighthouses or oil platforms. Birds of prey are diurnal because, although their eyes are large, they are optimised to give maximum spatial resolution rather than light gathering, so they also do not function well in poor light. Many birds have an asymmetry in the eye's structure which enables them to keep the horizon and a significant part of the ground in focus simultaneously. The cost of this adaptation is that they have myopia in the lower part of their visual field.

Birds with relatively large eyes compared to their body mass, such as common redstarts and European robins sing earlier at dawn than birds of the same size and smaller body mass. However, if birds have the same eye size but different body masses, the larger species sings later than the smaller. This may be because the smaller bird has to start the day earlier because of weight loss overnight. Overnight weight loss for small birds is typically 5-10% and may be over 15% on cold winter nights. In one study, robins put on more mass in their dusk feeding when nights were cold.

Nocturnal birds have eyes optimised for visual sensitivity, with large corneas relative to the eye's length, whereas diurnal birds have longer eyes relative to the corneal diameter to give greater visual acuity. Information about the activities of extinct species can be deduced from measurements of the sclerotic ring and orbit depth. For the latter measurement to be made, the fossil must have retained its three-dimensional shape, so activity pattern cannot be determined with confidence from flattened specimens like *Archaeopteryx*, which has a complete sclerotic ring but no orbit depth measurement.

Anatomy of the Eye

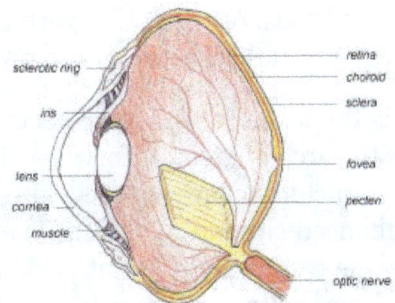

Anatomy of the avian eye

The main structures of the bird eye are similar to those of other vertebrates. The outer layer of the eye consists of the transparent cornea at the front, and two layers of sclera — a tough white collagen fibre layer which surrounds the rest of the eye and supports and protects the eye as a whole. The eye is divided internally by the lens into two main segments: the anterior segment and the posterior segment. The anterior chamber is filled with a watery fluid called the aqueous humour, and the posterior chamber contains the vitreous humour, a clear jelly-like substance.

The lens is a transparent convex or 'lens' shaped body with a harder outer layer and a softer inner layer. It focuses the light on the retina. The shape of the lens can be altered by ciliary muscles which are directly attached to lens capsule by means of the zonular fibres. In addition to these muscles, some birds also have a second set, Crampton's muscles, that can change the shape of the cornea, thus giving birds a greater range of accommodation than is possible for mammals. This accommodation can be rapid in some diving water birds such as in the mergansers. The iris is a coloured muscularly operated diaphragm in front of the lens which controls the amount of light entering the eye. At the centre of the iris is the pupil, the variable circular area through which the light passes into the eye.

Hummingbirds are amongst the many birds with two foveae

The retina is a relatively smooth curved multi-layered structure containing the photosensitive rod and cone cells with the associated neurons and blood vessels. The density of the photoreceptors is critical in determining the maximum attainable visual acuity. Humans have about 200,000 receptors per mm², but the house sparrow has 400,000 and the common buzzard 1,000,000. The photoreceptors are not all individually connected to the optic nerve, and the ratio of nerve ganglia to receptors is important in determining resolution. This is very high for birds; the white wagtail has 100,000 ganglion cells to 120,000 photoreceptors.

Rods are more sensitive to light, but give no colour information, whereas the less sensitive cones enable colour vision. In diurnal birds, 80% of the receptors may be cones (90% in some swifts) whereas nocturnal owls have almost all rods. As with other vertebrates except placental mammals, some of the cones may be double cones. These can amount to 50% of all cones in some species.

Towards the centre of the retina is the fovea (or the less specialised, area centralis) which has a greater density of receptors and is the area of greatest forward visual acuity, i.e. sharpest, clearest detection of objects. In 54% of birds, including birds of prey, kingfishers, hummingbirds and swallows, there is second fovea for enhanced sideways viewing. The optic nerve is a bundle of nerve fibres which carry messages from the eye to the relevant parts of the brain and vice versa. Like mammals, birds have a small blind spot without photoreceptors at the optic disc, under which the optic nerve and blood vessels join the eye.

The pecten is a poorly understood body consisting of folded tissue which projects from the retina. It is well supplied with blood vessels and appears to keep the retina supplied with nutrients, and may also shade the retina from dazzling light or aid in detecting moving objects. Pecten oculi is abundantly filled with melanin granules which have been proposed to absorb stray light entering the bird eye to reduce background glare. Slight warming of pecten oculi due to absorption of light by melanin granules has been proposed enhance metabolic rate of pecten that is suggested to help increase secretion of nutrients into vitreous, eventually to be absorbed by avascular retina of birds for improved nutrition. Exra-high enzymic activity of alkaline phosphatase in pecten oculi has been proposed to support high secretory activity of pecten to supplement nutrition of retina.

The choroid is a layer situated behind the retina which contains many small arteries and veins. These provide arterial blood to the retina and drain venous blood. The choroid contains melanin, a pigment which gives the inner eye its dark colour, helping to prevent disruptive reflections.

Light Perception

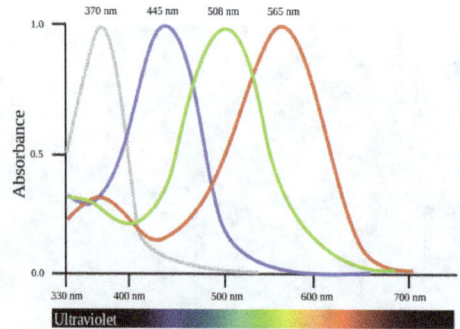

The four pigments in estrildid finches' cones extend the range of colour vision into the ultraviolet

There are two sorts of light receptors in a bird's eye, rods and cones. Rods, which contain the visual pigment rhodopsin are better for night vision because they are sensitive to small quantities of light. Cones detect specific colours (or wavelengths) of light, so they are more important to colour-orientated animals such as birds. Most birds are tetrachromatic, possessing four types of cone cells each with a distinctive maximal absorption peak. In some birds, the maximal absorption peak of the cone cell responsible for the shortest wavelength extends to the ultraviolet (UV) range, making them UV-sensitive. In addition to that, the cones at bird's retina are arranged in a characteristic form of spatial distribution, known as hyperuniform distribution, which maximizes its light and color absorption. This form of spatial distributions are only observed as a result of some optimization process, which in this case can be described in terms of bird's evolutionary history.

The four spectrally distinct cone pigments are derived from the protein opsin, linked to a small molecule called retinal, which is closely related to vitamin A. When the pigment absorbs light the retinal changes shape and alters the membrane potential of the cone cell affecting neurons in the ganglia layer of the retina. Each neuron in the ganglion layer may process information from a number of photoreceptor cells, and may in turn trigger a nerve impulse to relay information along the optic nerve for further processing in specialised visual centres in the brain. The more intense a light, the more photons are absorbed by the visual pigments; the greater the excitation of each cone, and the brighter the light appears.

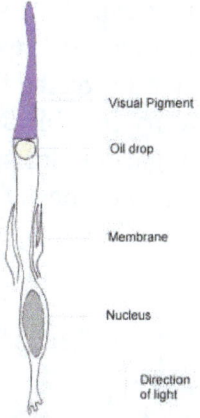

Visual Pigment

Oil drop

Membrane

Nucleus

Direction
of light

Diagram of a bird cone cell

By far the most abundant cone pigment in every bird species examined is the long-wavelength form of iodopsin, which absorbs at wavelengths near 570 nm. This is roughly the spectral region occupied by the red- and green-sensitive pigments in the primate retina, and this visual pigment dominates the colour sensitivity of birds. In penguins, this pigment appears to have shifted its absorption peak to 543 nm, presumably an adaptation to a blue aquatic environment.

The information conveyed by a single cone is limited: by itself, the cell cannot tell the brain which wavelength of light caused its excitation. A visual pigment may absorb two wavelengths equally, but even though their photons are of different energies, the cone cannot tell them apart, because they both cause the retinal to change shape and thus trigger the same impulse. For the brain to see colour, it must compare the responses of two or more classes of cones containing different visual pigments, so the four pigments in birds give increased discrimination.

Each cone of a bird or reptile contains a coloured oil droplet; these no longer exist in mammals. The droplets, which contain high concentrations of carotenoids, are placed so that light passes through them before reaching the visual pigment. They act as filters, removing some wavelengths and narrowing the absorption spectra of the pigments. This reduces the response overlap between pigments and increases the number of colours that a bird can discern. Six types of cone oil droplets have been identified; five of these have carotenoid mixtures that absorb at different wavelengths and intensities, and the sixth type has no pigments. The cone pigments with the lowest maximal absorption peak, including those that are UV-sensitive, possess the 'clear' or 'transparent' type of oil droplets with little spectral tuning effect.

The colours and distributions of retinal oil droplets vary considerably among species, and is more dependent on the ecological niche utilised (hunter, fisher, herbivore) than genetic relationships. As examples, diurnal hunters like the barn swallow and birds of prey have few coloured droplets, whereas the surface fishing common tern has a large number of red and yellow droplets in the dorsal retina. The evidence suggests that oil droplets respond to natural selection faster than the cone's visual pigments. Even within the range of wavelengths that are visible to humans, passerine birds can detect colour differences that humans do not register. This finer discrimination, together with the ability to see ultraviolet light, means that many species show sexual dichromatism that is visible to birds but not humans.

Migratory songbirds use the Earth's magnetic field, stars, the Sun, and other unknown cues to determine their migratory direction. An American study suggested that migratory Savannah sparrows used polarised light from an area of sky near the horizon to recalibrate their magnetic navigation system at both sunrise and sunset. This suggested that skylight polarisation patterns are the primary calibration reference for all migratory songbirds. However, it appears that birds may be responding to secondary indicators of the angle of polarisation, and may not be actually capable of directly detecting polarisation direction in the absence of these cues.

Ultraviolet Sensitivity

The common kestrel, like other raptorial birds, have a very low sensitivity to UV light.

There are two types of colour vision in birds: violet sensitive (VS) and ultraviolet sensitive (UVS). UVS birds have a visual pigment in the cones of their retinas that absorbs UV light, allowing them to see the ultraviolet portion of the spectrum. The major clades of birds that have UVS vision are Palaeognathae (ratites and tinamous), Charadriiformes (shorebirds, gulls, and alcids), Trogoniformes (trogons), Psittaciformes (parrots), and Passeriformes (perching birds, representing more than half of all avian species).

UVS vision can be useful for courtship. Many birds show plumage patterns in ultraviolet that are invisible to the human eye; some birds whose sexes appear similar to the naked eye are distinguished by the presence of ultraviolet reflective patches on their feathers. Male blue tits have an ultraviolet reflective crown patch which is displayed in courtship by posturing and raising of their nape feathers. Male blue grosbeaks with the brightest and most UV-shifted blue in their plumage are larger, hold the most extensive territories with abundant prey, and feed their offspring more frequently than other males do.

The bill's appearance is important in the interactions of the blackbird. Although the UV component seems unimportant in interactions between territory-holding males, where the degree of orange is the main factor, the female responds more strongly to males with bills with good UV-reflectiveness.

UVS vision may also give birds an advantage in foraging for food. The waxy surfaces of many fruits and berries reflect UV light that might advertise their presence. Common kestrels may be able to locate the trails of voles visually. These small rodents lay scent trails of urine and faeces that could reflect UV light, making them visible to the kestrels, particularly in the spring before the scent marks are covered by vegetation. However, this view has been challenged by the finding of low UV sensitivity in raptors and weak UV reflection of mammal urine.

While birds are not unique in their ability to perceive ultraviolet light (insects, reptiles, and crustaceans have UVS vision as well), some predators of UVS birds cannot see ultraviolet light. This raises the possibility that ultraviolet vision gives birds a channel in which they can privately signal, thereby remaining inconspicuous to predators. However, recent evidence does not appear to support this hypothesis.

Perception

Contrast Sensitivity

Contrast is defined as the difference in brightness between two stimuli, divided by the sum of brightness of both stimuli. Contrast sensitivity is the inverse of the smallest contrast that can be detected, a contrast sensitivity of 100 means that the smallest contrast that can be detected is 1%. Birds have comparably low contrast sensitivity to mammals. Humans have been shown to detect contrasts as low as 0.5-1%, whereas most birds tested have require ca 10% contrast to show a behavioural response. A contrast sensitivity function describes an animal's ability to resolve detect the contrast of patterns of different spatial frequency (detail). For stationary viewing experiments the contrast sensitivity is highest for a medium spatial frequency and lower for higher and lower spatial frequencies.

Movement

Birds can resolve rapid movements better than humans, for whom flickering at a rate greater than 50 light pulse cycles per second appears as continuous movement. Humans cannot therefore distinguish individual flashes of a fluorescent light bulb oscillating at 60 light pulse cycles per second, but budgerigars and chickens have flicker or light pulse cycles per second thresholds of more than 100 light pulse cycles per second which is safe to strictly say not Hertz A Cooper's hawk can pursue agile prey through woodland and avoid branches and other objects at high speed; to humans such a chase would appear as a blur.

A red kite flying at a bird feeding station in Scotland

Birds can also detect slow moving objects. The movement of the sun and the constellations across the sky is imperceptible to humans, but detected by birds. The ability to detect these movements allows migrating birds to properly orient themselves.

To obtain steady images while flying or when perched on a swaying branch, birds hold the head as steady as possible with compensating reflexes. Maintaining a steady image is especially relevant for birds of prey.

Edges and Shapes

When an object is partially blocked by another, humans unconsciously tend to make up for it and complete the shapes. It has however been demonstrated that pigeons do not complete occluded shapes. A study based on altering the grey level of a perch that was coloured differently from the background showed that budgerigars do not detect edges based on colours.

Magnetic Fields

The perception of magnetic fields by migratory birds has been suggested to be light dependent. Birds move their head to detect the orientation of the magnetic field, and studies on the neural pathways have suggested that birds may be able to "see" the magnetic fields. The right eye of a migratory bird contains photoreceptive proteins called cryptochromes. Light excites these molecules to produce unpaired electrons that interact with the Earth's magnetic field, thus providing directional information.

Variations Across Bird Groups

Diurnal Birds of Prey

The visual ability of birds of prey is legendary, and the keenness of their eyesight is due to a variety of factors. Raptors have large eyes for their size, 1.4 times greater than the average for birds of the same weight, and the eye is tube-shaped to produce a larger retinal image. The resolving power of an eye depends both on the optics, large eyes with large appertures suffers less from diffraction and can have larger retinal images due to a long focal length, and on the density of receptor spacing.

The retina has a large number of receptors per square millimeter, which determines the degree of visual acuity. The more receptors an animal has, the higher its ability to distinguish individual objects at a distance, especially when, as in raptors, each receptor is typically attached to a single ganglion. Many raptors have foveas with far more rods and cones than the human fovea (65,000/mm² in American kestrel, 38,000 in humans) and this provides these birds with spectacular long distance vision. It is proposed that the shape of the deep central fovea of raptors can create a tele-photo optical system, increasing the size of the retinal image in the fovea and thereby increasing the spatial resolution. Behavioural studies show that some large eyed raptors (Wedge-tailed eagle, Old world vultures) and have ca 2 times higher spatial resolution than humans, but many medium and small sized raptors have comparable or lower spatial resolution.

"Hawk-eyed" is a byword for visual acuity

The forward-facing eyes of a bird of prey give binocular vision, which is assisted by a double fovea. The raptor's adaptations for optimum visual resolution (an American kestrel can see a 2–mm insect from the top of an 18–m tree) has a disadvantage in that its vision is poor in low light level, and it must roost at night. Raptors may have to pursue mobile prey in the lower part of their visual field, and therefore do not have the lower field myopia adaptation demonstrated by many other birds. Scavenging birds like vultures do not need such sharp vision, so a condor has only a single fovea with about 35,000 receptors mm². Vultures, however have high physiological activity of many important enzymes to suit their distant clarity of vision.

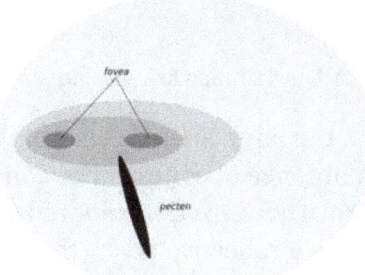

Each retina of the black-chested buzzard-eagle has two foveae

Like other birds investigated raptors do also have coloured oil droplets in their cones. The generally brown, grey and white plumage of this group, and the absence of colour displays in courtship suggests that colour is relatively unimportant to these birds.

In most raptors a prominent eye ridge and its feathers extends above and in front of the eye. This "eyebrow" gives birds of prey their distinctive stare. The ridge physically protects the eye from wind, dust, and debris and shields it from excessive glare. The osprey lacks this ridge, although the arrangement of the feathers above its eyes serves a similar function; it also possesses dark feathers in front of the eye which probably serve to reduce the glare from the water surface when the bird is hunting for its staple diet of fish.

Nocturnal Birds

A powerful owl photographed at night showing reflective tapeta lucida

Owls have very large eyes for their size, 2.2 times greater than the average for birds of the same weight, and positioned at the front of the head. The eyes have a field overlap of 50–70%, giving better binocular vision than for diurnal birds of prey (overlap 30–50%). The tawny owl's retina has about 56,000 light-sensitive rods per square millimetre (36 million per square inch); although earlier claims that it could see in the infrared part of the spectrum have been dismissed.

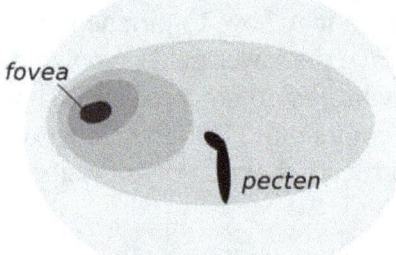

Each owl's retina has a single fovea

Adaptations to night vision include the large size of the eye, its tubular shape, large numbers of closely packed retinal rods, and an absence of cones, since cone cells are not sensitive enough for a low-photon nighttime environment. There are few coloured oil droplets, which would reduce the light intensity, but the retina contains a reflective layer, the tapetum lucidum. This increases the amount of light each photosensitive cell receives, allowing the bird to see better in low light conditions. Owls normally have only one fovea, and that is poorly developed except in diurnal hunters like the short-eared owl.

Besides owls, bat hawks, frogmouths and nightjars also display good night vision. Some bird species nest deep in cave systems which are too dark for vision, and find their way to the nest with a simple form of echolocation. The oilbird is the only nocturnal bird to echolocate, but several *Aero-dramus* swiftlets also utilise this technique, with one species, Atiu swiftlet, also using echolocation outside its caves.

Water Birds

Terns have coloured oil droplets in the cones of the eye to improve distance vision

Seabirds such as terns and gulls that feed at the surface or plunge for food have red oil droplets in the cones of their retinas. This improves contrast and sharpens distance vision, especially in hazy conditions. Birds that have to look through an air/water interface have more deeply coloured carotenoid pigments in the oil droplets than other species.

This helps them to locate shoals of fish, although it is uncertain whether they are sighting the phytoplankton on which the fish feed, or other feeding birds.

Birds that fish by stealth from above the water have to correct for refraction particularly when the fish are observed at an angle. Reef herons and little egrets appear to be able to make the corrections needed when capturing fish and are more successful in catching fish when strikes are made at an acute angle and this higher success may be due to the inability of the fish to detect their predators. Other studies indicate that egrets work within a preferred angle of strike and that the probability of misses increase when the angle becomes too far from the vertical leading to an increased difference between the apparent and real depth of prey.

Birds that pursue fish under water like auks and divers have far fewer red oil droplets, but they have special flexible lenses and use the nictitating membrane as an additional lens. This allows greater optical accommodation for good vision in air and water. Cormorants have a greater range of visual accommodation, at 50 dioptres, than any other bird, but the kingfishers are considered to have the best all-round (air and water) vision.

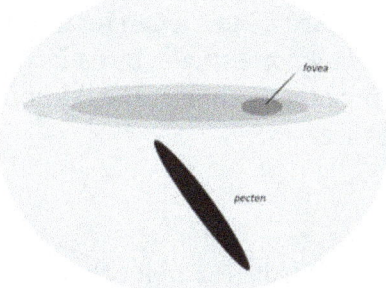

Each retina of the Manx shearwater has one fovea and an elongated strip of high photoreceptor density

Tubenosed seabirds, which come ashore only to breed and spend most of their life wandering close to the surface of the oceans, have a long narrow area of visual sensitivity on the retina This region, the *area giganto cellularis*, has been found in the Manx shearwater, Kerguelen petrel, great shearwater, broad-billed prion and common diving-petrel. It is characterised by the presence of ganglion cells which are regularly arrayed and larger than those found in the rest of the retina, and morphologically appear similar to the cells of the retina in cats. The location and cellular morphology of this novel area suggests a function in the detection of items in a small binocular field projecting below and around the bill. It is not concerned primarily with high spatial resolution, but may assist in the detection of prey near the sea surface as a bird flies low over it.

The Manx shearwater, like many other seabirds, visits its breeding colonies at night to reduce the chances of attack by aerial predators. Two aspects of its optical structure suggest that the eye of this species is adapted to vision at night. In the shearwater's eyes the lens does most of the bending of light necessary to produce a focused image on the retina. The cornea, the outer covering of the eye, is relative flat and so of low refractive power. In a diurnal bird like the pigeon, the reverse is true; the cornea is highly curved and is the principal refractive component. The ratio of refraction by the lens to that by the cornea is 1.6 for the shearwater and 0.4 for the pigeon; the figure for the shearwater is consistent with that for a range of nocturnal birds and mammals.

The shorter focal length of shearwater eyes give them a smaller, but brighter, image than is the case for pigeons, so the latter has sharper daytime vision. Although the Manx shearwater has adaptations for night vision, the effect is small, and it is likely that these birds also use smell and hearing to locate their nests.

It used to be thought that penguins were far-sighted on land. Although the cornea is flat and adapted to swimming underwater, the lens is very strong and can compensate for the reduced corneal focusing when out of water. Almost the opposite solution is used by the hooded merganser which can bulge part of the lens through the iris when submerged.

Avian Immune System

The avian immune system refers to the system of biological structures and cellular processes that protects birds from disease.

The avian immune system resembles that of mammals since both evolved from a common reptilian ancestor and have inherited many commonalities. They have also developed a number of different strategies that are unique to birds. Most avian immunology research has been carried out on the domestic chicken, *Gallus gallus domesticus*. Birds have lymphoid tissues, B cells, T cells, cytokines and chemokines like many other animals. In addition, they can also have tumours, immune deficiency and autoimmune diseases.

Overview

The physiology and immune system of birds resembles that of other animals. The lymphomyeloid tissues develop from epithelial or mesenchymal anlages that are full of haematopoetic cells. The bursa fabricus, thymus, spleen and lymph nodes all develop when haematopoetic stem cells enter the bursal or thymic anlages and become competent B and T cells. The avian immune system is divided into two types of immunity, the innate and adaptive ones. The innate immune system includes physical and chemical barriers, blood proteins and phagocytic cells. In addition, complement serum proteins, which are a part of the innate immune system, work with antibodies to lyse target cell. Adaptive immunity, on the other hand, kicks in when the innate system fails to stop invading pathogens. The adaptive response includes targeted recognition of specific molecular features on the surface of the pathogen. Birds, like other animals, have B cells, T cells and humoral immunity as part of their adaptive response.

Structure

Various bird organs function to differentiate avian immune cells: the thymus, Bursa of Fabricius and bone marrow are primary avian lymphoid organs whereas the spleen, mucosal associated lymphoid tissues (MALT), , germinal centers, and diffuse lymphoid tissues are secondary lymphoid organs. Birds do not have lymph nodes. The thymus, where T cells develop, is located in the neck of birds. The Bursa of Fabricius is an organ that is unique to birds and is the only site for B cell differentiation and maturation. Located in the rump of birds, this organ is full of stem cells and very active in young birds but atrophies after six months. Bronchial associated lymphoid tissue (BALT) and gut associated lymphoid tissue (GALT) are found along the bronchus and intestines, respectively. In the avian respiratory system, there are heterophils, which are an important part of bird immunity. Within the head, there is head associated lymphoid tissues (HALT) that contain the Harderian gland, lacrimal gland and other structures in the larynx or nasopharynx. The Harderian gland is located behind the eyeballs and is the major component of HALT. It contains a large number of plasma cells and is the main secretory body of antibodies. Alongside these primary and secondary lymphoid organs, there is also the lymphatic circulatory system of vessels and capillaries that communicate with the blood supply and transport the lymph fluid throughout the bird's body.

T cells

The antigen recognition by T cells is a remarkable process dependent on the T cell receptor (TCR). The TCR is randomly generated and thus has extensive diversity in the peptides-MHC complexes it can recognize. Using monoclonal antibodies that are specific for chicken T cell surface antigens, the development of T cells in birds is studied. The differentiation pathways, functional processes and molecules of T cells are highly conserved in birds. However, there are some novel features of T cells that are unique to birds. These include a new lineage of cytoplasmic CD3+ lymphoid cells (TCR0 cells) and a T cell sublineage that expresses a different receptor isotypes (TCR3) generated exclusively in the thymus. Homologues of the mammalian gamma, delta and alpha beta TCR (TCR1 and TCR2) are found in birds. However, a third TCR, called TCR3, has been found in avian T cell populations that lack both TCR1 and TCR2. These were found on all CD3+ T cells and were either CD4+ or CD8+. This subset of T cells, as others, develops in the thymus and gets seeded throughout the body with the exception of the intestines. The pattern of accessory molecules expressed by avian T cells resembles mammalian α/β T cells. High CD8 expression precedes the dual expression of CD4 and CD8 but following clonal selection and expansion, avian T cells cease to express either CD4 or CD8.

B cells

The central organ for B cell development in birds is the Bursa of Fabricius. The function of the bursa was discovered when it was surgically removed from neonatal chicks and this led to an impaired antibody response to *Salmonella typhimurium*. It is now clear that the bursa is the primary site of B cell lymphopoeisis and that avian B cell development has some unique properties compared to human or mouse models. Almost all the B cell progenitors in the bursa of 4-day-old chickens express IgM on their cell surface. Studies have shown that B cells of 4 – 8 week old birds are derived from 2 – 4 allotypically committed precursor cells in each follicle. Bursal follicles are colonized by 2-5 pre-bursal stem cells and these undergo extensive proliferation after they are committed to an allotype. Expression of IgM is controlled by a biological clock as opposed to the bursal microenvironment. Moreover, the source of all B cells in adult birds was determined to be a population of self-renewing sIg+ B cells.

Development

In studying the development of the avian immune system, the embryo offers several advantages such as the availability of many embryos at precise stages of development and distinct B and T cell systems. Each population differentiates from a primary lymphoid organ: T cells in the thymus and B cells in the Bursa of Fabricius. Research has found that early feeding of hydrated nutritional supplements in chickens heavily affects the immune system development. This is often measured by weight of the Bursa of Fabricius, improved resistance to disease and earlier appearance of IgA. Unlike other animals, newly hatched chicks are born with an incomplete immune system. Here, the amniotic fluid and yolk of the egg contain the maternal immunity to be passed on to the hatchling. Swallowing of the amniotic fluid during hatching confers immunity to these chicks until their immune system develops fully. In the first six weeks of the bird's life, continuous gene conversion in the bursa completes the immune system. Upon hatch, birds do not have a library of genetic information for B cells to use for antibody production. Instead, B cells mature in the bursa during the first six weeks and then go on to seed other organs of the immune system. As a result, birds are highly susceptible to pathogens in the first few weeks after hatching. Research found that T cells from mature chickens proliferated extensively and produced high levels of IL-2 and other cytokines. On the other hand, T cells from 24 hour-old chickens failed to proliferate and could not secrete cytokines. Gene conversion within the bursa leads to the development of antibodies that are diverse in their recognition ability. Mammalian V, D and J gene segments allow for many combinations and therefore, yield a vast repertoire of antibodies. However, birds have only a single functional copy of the V_L and J_L genes for the Ig light chain and a single functional copy of the V_H and J_H heavy chain genes. This results in a low diversity from gene rearrangements of Ig heavy and light chains. However, clusters of pseudogenes upstream of the heavy and light gene Ig loci take part in somatic gene conversion – a process where pseudogenes replace the V_H and V_L genes. This diversifies the repertoire of bird antibodies.

Avian Innate Immune System

Little is known about the innate immune system of birds. Most research has been focused on chickens due to the increased threat of viral diseases within the poultry population. The innate immune response is known to be essential for viral infection and as a result, the publication of the full chicken genome sequence is a source for identifying possible adjuvants and immunity genes.

Unique Features

Transfer of Maternal Immunity

Avian immunity begins to develop at the end of embryonic life but the majority of early immunity is obtained via passive acquisition of maternal antibodies. Such antibodies are found within the egg when it is laid and originated from the yolk of the egg. Kramer and Cho have shown immunoglobulins in both the egg white and in the embryo. Maternal IgA and IgM get transferred to the egg as it passes down the oviduct.

TTP

An important element of immune systems in various animals is the protein tristetraprolin (TTP). This plays a key anti-inflammatory role by regulating TNFα. Mouse models with TTP knockouts result in chronic and often deadly inflammation when exposed to small amounts of pathogen-associated molecular patterns (PAMPs). However, TTP and its homologs is altogether absent from birds. Avian genomes have been searched for similar sequences to TTP and bird cell lines have been exposed to foreign proteins and bacteria molecules known to stimulate TTP production but no evidence of TTP has been found. The missing protein poses a very different immune response regulation in birds as opposed to mammals, reptiles and amphibians.

Organs

The avian T cell population, like that of mammals develops in the thymus. However, the thymus in birds is a paired organ composed of many separated lobes of ovoid tissue in the neck. These are close to the vagus nerve and the jugular vein and are most active in young hatchlings. It is postulated that this organ is linked to erythropoeitic function and closely associated with the avian breeding cycle. The removal of the thymic lobes has been correlated to birds rejecting allogeneic skingrafts and delayed skin reactions.

The bursa of Fabricus is a globular or spherical lymphoepithelial organ. The inner surface is littered with folds, which resemble Peyer's patches in mammals and obscure the lumen. Its growth is correlated with the rapid body growth. Interestingly, it regresses and disappears about the time of sexual maturity. The bursa, as studied through bursectomy at different developmental stages, indicates sequential development of IgG, IgM and IgA. The secondary (peripheral) lymphoid tissue also includes unique lymphoid nodules in the digestive tract and solitary nodules scattered throughout the body, a characteristic of avian species. Meanwhile, lymph nodes only occur in some water, marsh and shore species.

Diseases

Control of infectious disease is essential for the production of healthy poultry flocks. Vaccination programs have been used extensively in North American factory farming methods to induce avian immune responses against bird pathogens. These include Marek's Disease, Duck Hepatitis Virus,

Chicken Anemia Virus, Turkeypox, Fowlpox and others. Bird immunity is reliant on a complex network of cell types and soluble factors that must properly function in order for large commercial poultry flocks to survive.

Infectious bursal disease virus and chicken anemia are ubiquitous and have increased interest in combatting avian pathogens. Parasites of birds are another emerging concern since the crowded nature of poultry farms facilitates easy spreading.

Immunosuppressive Diseases

Several immunosuppressive agents are encountered by birds including viruses, bacteria, parasites, toxins, mycotoxins, chemicals and drugs. The most common immunosuppressive viruses are Infectious Bursal Disease Virus (IBDV), Avian Leukosis, Marek's Disease (MD) and Hemorrhagic Enteritis Virus (HEV). Concurrent immunosuppressive infections are an emerging concern in the poultry industry whereby early infection with IBDV causes the MD virus to come out of dormancy and contribute to active disease. New studies show that stress is the number one cause of immunosuppression in birds. Stressors leave birds more susceptible to infectious agents and therefore, new poultry management guidelines need to be endorsed.

Birds as Vectors

The migratory nature of birds poses a distinct danger for the spreading of diseases. Without being affected by the infectious agent, birds can act as vectors in spreading psittacosis, salmonellosis, campylobacteriosis, mycobacteriosis, avian influenza, giardiasis and cryptosporidiosis. These zoonotic diseases can be transmitted to humans. In the case of avian influenza (H5N1 strain), water birds can be infected with the low pathogenic form or the high pathogenic form. The former induces mild symptoms such as a drop in egg production, ruffled feathers and mild effects on the avian respiratory tract. The highly pathogenic form spreads much more rapidly and can infect multiple tissues and organs. Massive internal bleeding and hemorrhaging follow and this has earned the H5N1 virus the moniker "chicken ebola."

Tumours

Much like other animals, birds are prone to cancers and tumours. This refers to the abnormal growth of cells in a tissue or organ that can be either malignant or benign. Internal cancers can occur in the kidneys, liver, stomach, ovary, muscles or bone. Squamous cell carcinoma is a form of skin cancer that birds obtain, manifesting on the wing tips, toes, and around the beak and eyes. The cause is believed to be high exposure to UV rays. Additionally, a cancer of the connective tissue, known as fibrosarcoma, is often seen in the leg or wing. This occurs in many parrot species, cockatiels, macaws and budgerigars. Treatment options include amputation and surgery.

References

- Scott, Graham R. (2011). "Commentary: Elevated performance: the unique physiology of birds that fly at high altitudes". Journal of Experimental Biology. 214: 2455–2462. doi:10.1242/jeb.052548

- Proctor, N. S. & Lynch, P. J. (1998) Manual of Ornithology: Avian Structure & Function. Yale University Press. ISBN 0300076193

- Gier, H. T. (1952). "The air sacs of the loon" (PDF). Auk. American Ornithologists' Union. 69: 40–49. doi:10.2307/4081291. Retrieved 2014-01-21

- Stettenheim Peter R (2000). "The Integumentary Morphology of Modern Birds—An Overview". American Zoologist. 40 (4): 461–477. doi:10.1093/icb/40.4.461

- Zaher, Mostafa (2012). "Anatomical, histological and histochemical adaptations of the avian alimentary canal to their food habits: I-Coturnix coturnix" (PDF). Life Science Journal. 9: 253–275

- Calder, William A. (1996). Size, Function, and Life History. Mineola, New York: Courier Dove Publications. p. 91. ISBN 978-0-486-69191-6

- Ritchson, G. "BIO 554/754 – Ornithology: Avian respiration". Department of Biological Sciences, Eastern Kentucky University. Retrieved 2009-04-23

- Herrera, A. M; S. G. Shuster; C. L. Perriton; M. J. Cohn (2013). "Developmental Basis of Phallus Reduction during Bird Evolution". Current Biology. 23 (12): 1065–1074. PMID 23746636. doi:10.1016/j.cub.2013.04.062

- Kinsky, FC (1971). "The consistent presence of paired ovaries in the Kiwi(Apteryx) with some discussion of this condition in other birds". Journal of Ornithology. 112 (3): 334–357. doi:10.1007/BF01640692

- Lynch, Wayne; Lynch, photographs by Wayne (2007). Owls of the United States and Canada : a complete guide to their biology and behavior. Baltimore: Johns Hopkins University Press. p. 151. ISBN 0-8018-8687-2

- Gier, H. T. (1952). "The air sacs of the loon" (PDF). The Auk. American Ornithologists' Union. 69 (1): 40–49. doi:10.2307/4081291. Retrieved 21 January 2014

- Hosken, D.J.; P. Stockley (2004). "Sexual selection and genital evolution". Trends in Ecology & Evolution. 19: 87–93. doi:10.1016/j.tree.2003.11.012

- Mills, Robert (March 1994). "Applied comparative anatomy of the avian middle ear" (PDF). Journal of the Royal Society of Medicine. 87: 155. Retrieved 17 March 2017

- Campbell JA, Lamar WW. 2004. The Venomous Reptiles of the Western Hemisphere. 2 volumes. Comstock Publishing Associates, Ithaca and London. 870 pp. 1500 plates. ISBN 0-8014-4141-

- Whitfield, John (10 March 2000). "Off to a flying jump-start : Nature News". Nature Publishing Group. doi:10.1038/news000316-1. Retrieved 2014-01-17

- Bonser RH & Mark S Witter (1993). "Indentation hardness of the bill keratin of the European Starling" (PDF). The Condor. 95: 736–738. doi:10.2307/1369622

- Jones, E.K.M. and Prescott, N.B., (2000). Visual cues used in the choice of mate by fowl and their potential importance for the breeder industry. World's Poultry Science Journal, 56: 127-138. doi:10.1079/WPS20000010

- Pyle, Peter; Howell, Steve N. G.; Yunick, Robert P.; DeSante, David F. (1987). Identification Guide to North America Passerines. Bolinas, CA: Slate Creek Press. pp. 6–7. ISBN 0-9618940-0-8

- Hieronymus, Tobin L.; Witmer, Lawrence M. (2010). "Homology and Evolution of Avian Compound Rhamphothecae". The Auk. 127 (3): 590–604. doi:10.1525/auk.2010.09122

- Amerson, A. Binion (May 1967). "Incidence and Transfer of Rhinonyssidae (Acarina: Mesostigmata) in Sooty Terns (Sterna fuscata)". Journal of Medical Entomology. 4 (2): 197–9. PMID 6052126

- Schuetz, Justin G. (October 2005). "Reduced growth but not survival of chicks with altered gape patterns". Animal Behaviour. 70 (4): 839–848. ISSN 0003-3472. doi:10.1016/j.anbehav.2005.01.007

- "Bird Beaks: Anatomy, Care, and Diseases". Veterinary & Aquatic Services Department, Drs. Foster & Smith. Retrieved 16 April 2012

5

An Overview of Animal Locomotion

Animal locomotion is the term used for the movement of animals. Some of the methods used for movement by animals are running, jumping, flying, hopping and swimming. Terrestrial locomotion, arboreal locomotion and aquatic locomotion are also discussed. This chapter is an overview of the subject matter incorporating all the major aspects of animal locomotion.

Animal Locomotion

Animal locomotion, in ethology, is any of a variety of movements or methods that animals use to move from one place to another. Some modes of locomotion are (initially) self-propelled, e.g., running, swimming, jumping, flying, hopping, soaring and gliding. There are also many animal species that depend on their environment for transportation, a type of mobility called passive locomotion, e.g., sailing (some jellyfish), kiting (spiders) and rolling (some beetles and spiders).

A cheetah chasing prey. In its brief hunting sprints, it is the fastest of all land animals.

Animals move for a variety of reasons, such as to find food, a mate, a suitable microhabitat, or to escape predators. For many animals, the ability to move is essential for survival and, as a result, natural selection has shaped the locomotion methods and mechanisms used by moving organisms. For example, migratory animals that travel vast distances (such as the Arctic tern) typically have a locomotion mechanism that costs very little energy per unit distance, whereas non-migratory animals that must frequently move quickly to escape predators are likely to have energetically costly, but very fast, locomotion.

The anatomical structures that animals use for movement, including cilia, legs, wings, arms, fins, or tails, in various modes and through various media are sometimes referred to as *locomotory organs* or *locomotory structures*.

A bee in flight

Etymology

The term "locomotion" is formed in English from Latin *loco* "from a place" (ablative of *locus* "place") + *motio* "motion, a moving".

Locomotion in Different Media

Animals move through, or on, four types of environment: aquatic (in or on water), terrestrial (on ground or other surface, including arboreal, or tree-dwelling), fossorial (underground), and aerial (in the air). Many animals—for example semi-aquatic animals, and diving birds—regularly move through more than one type of medium. In some cases, the surface they move on facilitates their method of locomotion.

Aquatic

Swimming

Dolphins surfing

In water, staying afloat is possible using buoyancy. If an animal's body is less dense than water, it can stay afloat. This requires little energy to maintain a vertical position, but requires more energy for locomotion in the horizontal plane compared to less buoyant animals. The drag encountered in water is much greater than in air. Morphology is therefore important for efficient locomotion, which is in most cases essential for basic functions such as catching prey. A fusiform, torpedo-like body form is seen in many aquatic animals, though the mechanisms they use for locomotion are diverse.

The primary means by which fish generate thrust is by oscillating the body from side-to-side, the resulting wave motion ending at a large tail fin. Finer control, such as for slow movements, is often achieved with thrust from pectoral fins (or front limbs in marine mammals). Some fish, e.g. the spotted ratfish (*Hydrolagus colliei*) and batiform fish (electric rays, sawfishes, guitarfishes, skates and stingrays) use their pectoral fins as the primary means of locomotion, sometimes termed labriform swimming. Marine mammals oscillate their body in an up-and-down (dorso-ventral) direction. Other animals, e.g. penguins, diving ducks, move underwater in a manner which has been termed "aquatic flying". Some fish propel themselves without a wave motion of the body, as in the slow-moving seahorses and *Gymnotus*. Other animals, such as cephalopods, use jet propulsion to travel fast, taking in water then squirting it back out in an explosive burst. Other swimming animals may rely predominantly on their limbs, much as humans do when swimming. Though life on land originated from the seas, terrestrial animals have returned to an aquatic lifestyle on several occasions, such as the fully aquatic cetaceans, now very distinct from their terrestrial ancestors.

Dolphins sometimes ride on the bow waves created by boats or surf on naturally breaking waves.

Benthic

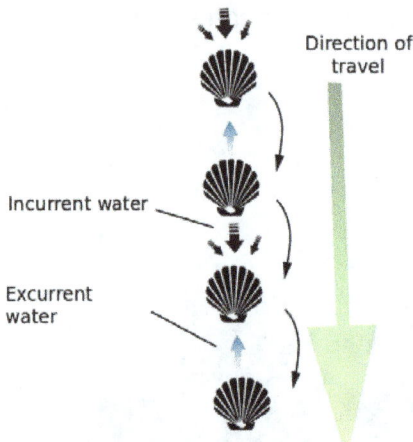

Scallop in jumping motion; these bivalves can also swim.

Benthic locomotion is movement by animals that live on, in, or near the bottom of aquatic environments. In the sea, many animals walk over the seabed. Echinoderms primarily use their tube feet to move about. The tube feet typically have a tip shaped like a suction pad that can create a vacuum through contraction of muscles. This, along with some stickiness from the secretion of mucus, provides adhesion. Waves of tube feet contractions and relaxations move along the adherent surface and the animal moves slowly along. Some sea urchins also use their spines for benthic locomotion.

Crabs typically walk sideways (a behaviour that gives us the word *crabwise*). This is because of the articulation of the legs, which makes a sidelong gait more efficient. However, some crabs walk forwards or backwards, including raninids, *Libinia emarginata* and *Mictyris platycheles*. Some crabs, notably the Portunidae and Matutidae, are also capable of swimming, the Portunidae especially so as their last pair of walking legs are flattened into swimming paddles.

A stomatopod, *Nannosquilla decemspinosa*, can escape by rolling itself into a self-propelled wheel and somersault backwards at a speed of 72 rpm. They can travel more than 2 m using this unusual method of locomotion.

Aquatic Surface

Velella moves by sailing

Velella, the by-the-wind sailor, is a cnidarian with no means of propulsion other than sailing. A small rigid sail projects into the air and catches the wind. *Velella* sails always align along the direction of the wind where the sail may act as an aerofoil, so that the animals tend to sail downwind at a small angle to the wind.

While larger animals such as ducks can move on water by floating, some small animals move across it without breaking through the surface. This surface locomotion takes advantage of the surface tension of water. Animals that move in such a way include the water strider. Water striders have legs that are hydrophobic, preventing them from interfering with the structure of water. Another form of locomotion (in which the surface layer is broken) is used by the basilisk lizard.

Aerial

Active Flight

Gravity is the primary obstacle to flight. Because it is impossible for any organism to have a density as low as that of air, flying animals must generate enough lift to ascend and remain airborne. One way to achieve this is with wings, which when moved through the air generate an upward lift force on the animal's body. Flying animals must be very light to achieve flight, the largest living flying animals being birds of around 20 kilograms. Other structural adaptations of flying animals include reduced and redistributed body weight, fusiform shape and powerful flight muscles; there may also be physiological adaptations. Active flight has independently evolved at least four times, in the insects, pterosaurs, birds, and bats. Insects were the first taxon to evolve flight, approximately 400 million years ago (mya), followed by pterosaurs approximately 220 mya, birds approximately 160 mya, then bats about 60 mya.

A pair of brimstone butterflies in flight. The female, above, is in fast forward flight with a small angle of attack; the male, below, is twisting his wings sharply upward to gain lift and fly up towards the female.

Gliding

Rather than active flight, some (semi-) arboreal animals reduce their rate of falling by gliding. Gliding is heavier-than-air flight without the use of thrust; the term "volplaning" also refers to this mode of flight in animals. This mode of flight involves flying a greater distance horizontally than vertically and therefore can be distinguished from a simple descent like a parachute. Gliding has evolved on more occasions than active flight. There are examples of gliding animals in several major taxonomic classes such as the invertebrates (e.g., gliding ants), reptiles (e.g., banded flying snake), amphibians (e.g., flying frog), mammals (e.g., sugar glider, squirrel glider).

Flying fish taking off

Some aquatic animals also regularly use gliding, for example, flying fish, octopus and squid. The flights of flying fish are typically around 50 meters (160 ft), though they can use updrafts at the leading edge of waves to cover distances of up to 400 m (1,300 ft). To glide upward out of the water, a flying fish moves its tail up to 70 times per second. Several oceanic squid, such as the Pacific flying squid, leap out of the water to escape predators, an adaptation similar to that of flying fish. Smaller squids fly in shoals, and have been observed to cover distances as long as 50 m. Small fins towards the back of the mantle help stabilize the motion of flight. They exit the water by expelling water out of their funnel, indeed some squid have been observed to continue jetting water while airborne providing thrust even after leaving the water. This may make flying squid the only animals with jet-propelled aerial locomotion. The neon flying squid has been observed to glide for distances over 30 m, at speeds of up to 11.2 m/s.

Soaring

Soaring birds can maintain flight without wing flapping, using rising air currents. Many gliding birds are able to "lock" their extended wings by means of a specialized tendon. Soaring birds may alternate glides with periods of soaring in rising air. Five principal types of lift are used: thermals, ridge lift, lee waves, convergences and dynamic soaring.

Examples of soaring flight by birds are the use of:

- Thermals and convergences by raptors such as vultures

- Ridge lift by gulls near cliffs

- Wave lift by migrating birds

- Dynamic effects near the surface of the sea by albatrosses

Ballooning

Ballooning is a method of locomotion used by spiders. Certain silk-producing arthropods, mostly small or young spiders, secrete a special light-weight gossamer silk for ballooning, sometimes traveling great distances at high altitude.

Terrestrial

Forms of locomotion on land include walking, running, hopping or jumping, dragging and crawling or slithering. Here friction and buoyancy are no longer an issue, but a strong skeletal and muscular framework are required in most terrestrial animals for structural support. Each step also requires much energy to overcome inertia, and animals can store elastic potential energy in their tendons to help overcome this. Balance is also required for movement on land. Human infants learn to crawl first before they are able to stand on two feet, which requires good coordination as well as physical development. Humans are bipedal animals, standing on two feet and keeping one on the ground at all times while walking. When running, only one foot is on the ground at any one time at most, and both leave the ground briefly. At higher speeds momentum helps keep the body upright, so more energy can be used in movement.

Jumping

Jumping (saltation) can be distinguished from running, galloping, and other gaits where the entire body is temporarily airborne by the relatively long duration of the aerial phase and high angle of initial launch. Many terrestrial animals use jumping (including hopping or leaping) to escape predators or catch prey—however, relatively few animals use this as a primary mode of locomotion. Those that do include the kangaroo and other macropods, rabbit, hare, jerboa, hopping mouse, and kangaroo rat. Kangaroo rats often leap 2 m and reportedly up to 2.75 m at speeds up to almost 3 m/s (6.7 mph). They can quickly change their direction between jumps. The rapid locomotion of the banner-tailed kangaroo rat may minimize energy cost and predation risk. Its use of a "move-freeze" mode may also make it less conspicuous to nocturnal predators. Frogs are, relative to their size, the best jumpers of all vertebrates. The Australian rocket frog, *Litoria nasuta*, can leap over 2 metres (6 ft 7 in), more than fifty times its body length.

Peristalsis

Other animals move in terrestrial habitats without the aid of legs. Earthworms crawl by a peristalsis, the same rhythmic contractions that propel food through the digestive tract.

Sliding

Due to its low coefficient of friction, ice provides the opportunity for other modes of locomotion. Penguins either waddle on their feet or slide on their bellies across the snow, a movement called *tobogganing*, which conserves energy while moving quickly. Some pinnipeds perform a similar behaviour called *sledding*.

Brachiation

Some animals are specialized for moving on non-horizontal surfaces. One common habitat for such climbing animals is in trees, for example the gibbon is specialized for arboreal movement, traveling rapidly by brachiation. Another case is animals like the snow leopard living on steep rock faces such as are found in mountains. Some light animals are able to climb up smooth sheer surfaces or hang upside down by adhesion. Many insects can do this, though much larger animals such as geckos can also perform similar feats.

Walking and Running

Species have different numbers of legs resulting in large differences in locomotion.

Modern birds, though classified as tetrapods, usually have only two functional legs, which some (e.g., ostrich, emu, kiwi) use as their primary, Bipedal, mode of locomotion. Few modern mammalian species are habitual bipeds whose normal method of locomotion is two-legged. These include the macropods, kangaroo rats and mice, springhare, hopping mice, pangolins and homininan apes. Bipedalism is rarely found outside terrestrial animals—though at least two types of octopus walk bipedally on the sea floor using two of their arms, so they can use the remaining arms to camouflage themselves as a mat of algae or floating coconut.

There are no three-legged animals—though some macropods, such as kangaroos, that alternate between resting their weight on their muscular tails and their two hind legs, could be looked at as an example of tripedal locomotion in animals.

AN EIFELIAN TETRAPOD TRACK AND ITS MAKER
Muz. PGI 1728.II.16 - the Zachełmie quarry, Poland
based on Niedźwiedzki et al. 2010

Animation of a Devonian tetrapod

Many familiar animals are quadrupedal, walking or running on four legs. A few birds use quadrupedal movement in some circumstances. For example, the shoebill sometimes uses its wings to right itself after lunging at prey. The newly hatched hoatzin bird has claws on its thumb and first finger enabling it to dexterously climb tree branches until its wings are strong enough for sustained flight. These claws are gone by the time the bird reaches adulthood.

A relatively few animals use five limbs for locomotion. Prehensile quadrupeds may use their tail to assist in locomotion and when grazing, the kangaroos and other macropods use their tail to propel themselves forward with the four legs used to maintain balance.

Insects generally walk with six legs—though some insects such as nymphalid butterflies do not use the front legs for walking.

Arachnids have eight legs. Most arachnids lack extensor muscles in the distal joints of their appendages. Spiders and whipscorpions extend their limbs hydraulically using the pressure of their hemolymph. Solifuges and some harvestmen extend their knees by the use of highly elastic thickenings in the joint cuticle. Scorpions, pseudoscorpions and some harvestmen have evolved muscles that extend two leg joints (the femur-patella and patella-tibia joints) at once.

The scorpion *Hadrurus arizonensis* walks by using two groups of legs (left 1, right 2, Left 3, Right 4 and Right 1, Left 2, Right 3, Left 4) in a reciprocating fashion. This alternating tetrapod coordination is used over all walking speeds.

Centipedes and millipedes have many sets of legs that move in metachronal rhythm. Some echinoderms locomote using the many tube feet on the underside of their arms. Although the tube feet resemble suction cups in appearance, the gripping action is a function of adhesive chemicals rather than suction. Other chemicals and relaxation of the ampullae allow for release from the substrate. The tube feet latch on to surfaces and move in a wave, with one arm section attaching to the surface as another releases. Some multi-armed, fast-moving starfish such as the sunflower seastar (*Pycnopodia helianthoides*) pull themselves along with some of their arms while letting others trail behind. Other starfish turn up the tips of their arms while moving, which exposes the sensory tube feet and eyespot to external stimuli. Most starfish cannot move quickly, a typical speed being that of the leather star (*Dermasterias imbricata*), which can manage just 15 cm (6 in) in a minute. Some burrowing species from the genera *Astropecten* and *Luidia* have points rather than suckers on their long tube feet and are capable of much more rapid motion, "gliding" across the ocean floor. The sand star (*Luidia foliolata*) can travel at a speed of 2.8 m (9 ft 2 in) per minute. Sunflower starfish are quick, efficient hunters, moving at a speed of 1 m/min (3.3 ft/min) using 15,000 tube feet.

Many animals temporarily change the number of legs they use for locomotion in different circumstances. For example, many quadrupedal animals switch to bipedalism to reach low-level browse on trees. The genus of *Basiliscus* are arboreal lizards that usually use quadrupedalism in the trees. When frightened, they can drop to water below and run across the surface on their hind limbs at about 1.5 m/s for a distance of approximately 4.5 metres (15 ft) before they sink to all fours and swim. They can also sustain themselves on all fours while "water-walking" to increase the distance travelled above the surface by about 1.3 metres. When cockroaches run rapidly, they rear up on their two hind legs like bipedal humans; this allows them to run at speeds up to 50 body lengths/s, equivalent to a "couple hundred miles per hour, if you scale up to the size of humans". When grazing, kangaroos use a form of pentapedalism (four legs plus the tail) but switch to hopping (bipedalism) when they wish to move at a greater speed.

Powered Cartwheeling

The Moroccan flic-flac spider (*Cebrennus rechenbergi*) uses a series of rapid, acrobatic flic-flac movements of its legs similar to those used by gymnasts, to actively propel itself off the ground, allowing it to move both down and uphill, even at a 40 percent incline. This behaviour is different than other huntsman spiders, such as *Carparachne aureoflava* from the Namib Desert, which uses passive cartwheeling as a form of locomotion. The flic-flac spider can reach speeds of up to 2 m/s using forward or back flips to evade threats.

Subterranean

Some animals move through solids such as soil by burrowing using peristalsis, as in earthworms, or other methods. In loose solids such as sand some animals, such as the golden mole, marsupial mole, and the pink fairy armadillo, are able to move more rapidly, 'swimming' through the loose substrate. Burrowing animals include moles, ground squirrels, naked mole-rats, tilefish, and mole crickets.

Arboreal Locomotion

A brachiating gibbon

Arboreal locomotion is the locomotion of animals in trees. Some animals may only scale trees occasionally, while others are exclusively arboreal. These habitats pose numerous mechanical challenges to animals moving through them, leading to a variety of anatomical, behavioural and ecological consequences as well as variations throughout different species. Furthermore, many of these same principles may be applied to climbing without trees, such as on rock piles or mountains. The earliest known tetrapod with specializations that adapted it for climbing trees, was *Suminia*, a synapsid of the late Permian, about 260 million years ago. Some invertebrate animals are exclusively arboreal in habitat, for example, the tree snail.

Brachiation (from *brachium*, Latin for "arm"), is a form of arboreal locomotion in which primates swing from tree limb to tree limb using only their arms. During brachiation, the body is alternately supported under each forelimb. This is the primary means of locomotion for the small gibbons and siamangs of southeast Asia. Some New World monkeys such as spider monkeys and muriquis are "semibrachiators" and move through the trees with a combination of leaping and brachiation. Some New World species also practice suspensory behaviors by using their prehensile tail, which acts as a fifth grasping hand.

Energetics

Animal locomotion requires energy to overcome various forces including friction, drag, inertia and gravity, although the influence of these depends on the circumstances. In terrestrial environments, gravity must be overcome whereas the drag of air has little influence. In aqueous environments, friction (or drag) becomes the major energetic challenge with gravity being less of an influence. Remaining in the aqueous environment, animals with natural buoyancy expend little energy to maintain a vertical position in a water column. Others naturally sink, and must spend energy to remain afloat. Drag is also an energetic influence in flight, and the aerodynamically efficient body shapes of flying birds indicate how they have evolved to cope with this. Limbless organisms moving on land must energetically overcome surface friction, however, they do not usually need to expend significant energy to counteract gravity.

Newton's third law of motion is widely used in the study of animal locomotion: if at rest, to move forwards an animal must push something backwards. Terrestrial animals must push the solid ground, swimming and flying animals must push against a fluid (either water or air). The effect of forces during locomotion on the design of the skeletal system is also important, as is the interaction between locomotion and muscle physiology, in determining how the structures and effectors of locomotion enable or limit animal movement. The energetics of locomotion involves the energy expenditure by animals in moving. Energy consumed in locomotion is not available for other efforts, so animals typically have evolved to use the minimum energy possible during movement. However, in the case of certain behaviors, such as locomotion to escape a predator, performance (such as speed or maneuverability) is more crucial, and such movements may be energetically expensive. Furthermore, animals may use energetically expensive methods of locomotion when environmental conditions (such as being within a burrow) preclude other modes.

The most common metric of energy use during locomotion is the net also termed "incremental" cost of transport, defined as the amount of energy (e.g., Joules) needed above baseline metabolic rate to move a given distance. For aerobic locomotion, most animals have a nearly constant cost of transport - moving a given distance requires the same caloric expenditure, regardless of speed.

This constancy is usually accomplished by changes in gait. The net cost of transport of swimming is lowest, followed by flight, with terrestrial limbed locomotion being the most expensive per unit distance. However, because of the speeds involved, flight requires the most energy per unit time. This does not mean that an animal that normally moves by running would be a more efficient swimmer; however, these comparisons assume an animal is specialized for that form of motion. Another consideration here is body mass—heavier animals, though using more total energy, require less energy *per unit mass* to move. Physiologists generally measure energy use by the amount of oxygen consumed, or the amount of carbon dioxide produced, in an animal's respiration. In terrestrial animals, the cost of transport is typically measured while they walk or run on a motorized treadmill, either wearing a mask to capture gas exchange or with the entire treadmill enclosed in a metabolic chamber. For small rodents, such as deer mice, the cost of transport has also been measured during voluntary wheel running.

Energetics is important for explaining the evolution of foraging economic decisions in organisms; for example, a study of the African honey bee, *A. m. scutellata*, has shown that honey bees may trade-off the high sucrose content of viscous nectar for the energetic benefits of warmer, less concentrated nectar, which also reduces their consumption and flight time.

Passive Locomotion

Passive locomotion in animals is a type of mobility in which the animal depends on their environment for transportation.

Hydrozoans

Physalia physalis

The Portuguese man o' war (*Physalia physalis*) lives at the surface of the ocean. The gas-filled bladder, or pneumatophore (sometimes called a "sail"), remains at the surface, while the remainder is submerged. Because the Portuguese man o' war has no means of propulsion, it is moved by a combination of winds, currents, and tides. The sail is equipped with a siphon. In the event of a surface attack, the sail can be deflated, allowing the organism to briefly submerge.

Arachnids

The wheel spider (*Carparachne aureoflava*) is a huntsman spider approximately 20 mm in size and native to the Namib Desert of Southern Africa. The spider escapes parasitic pompilid wasps by flipping onto its side and cartwheeling down sand dunes at speeds of up to 44 turns per second. If the spider is on a sloped dune, its rolling speed may be 1 metre per second.

A spider (usually limited to individuals of a small species), or spiderling after hatching, climbs as high as it can, stands on raised legs with its abdomen pointed upwards ("tiptoeing"), and then releases several silk threads from its spinnerets into the air. These form a triangle-shaped parachute that carries the spider on updrafts of winds, where even the slightest breeze transports it. The Earth's static electric field may also provide lift in windless conditions.

Insects

The larva of *Cicindela dorsalis*, the eastern beach tiger beetle, is notable for its ability to leap into the air, loop its body into a rotating wheel and roll along the sand at a high speed using wind to propel itself. If the wind is strong enough, the larva can cover up to 60 metres (200 ft) in this manner. This remarkable ability may have evolved to help the larva escape predators such as the tiphiid wasp *Methocha*.

Members of the largest subfamily of cuckoo wasps, Chrysidinae, are generally kleptoparasites, laying their eggs in host nests, where their larvae consume the host egg or larva while it is still young. Chrysidines are distinguished from the members of other subfamilies in that most have flattened or concave lower abdomens and can curl into a defensive ball when attacked by a potential host, a process known as conglobation. Protected by hard chitin in this position, they are expelled from the nest without injury and can search for a less hostile host.

Fleas can jump vertically up to 18 cm and horizontally up to 33 cm, however, although this form of locomotion is initiated by the flea, it has little control of the jump - they always jump in the same direction, with very little variation in the trajectory between individual jumps.

Crustaceans

Although stomatopods typically display the standard locomotion types as seen in true shrimp and lobsters, one species, *Nannosquilla decemspinosa*, has been observed flipping itself into a crude wheel. The species lives in shallow, sandy areas. At low tides, *N. decemspinosa* is often stranded by its short rear legs, which are sufficient for locomotion when the body is supported by water, but not on dry land. The mantis shrimp then performs a forward flip in an attempt to roll towards the next tide pool. *N. decemspinosa* has been observed to roll repeatedly for 2 metres (6.6 ft), but they typically travel less than 1 m (3.3 ft). Again, the animal initiates the movement but has little control during its locomotion.

Animal Transport

Some animals change location because they are attached to, or reside on, another animal or moving structure. This is arguably more accurately termed "animal transport".

Remoras

Remoras are a family (*Echeneidae*) of ray-finned fish. They grow to 30–90 cm (0.98–2.95 ft) long, and their distinctive first dorsal fins take the form of a modified oval, sucker-like organ with slat-like structures that open and close to create suction and take a firm hold against the skin of larger marine animals. By sliding backward, the remora can increase the suction, or it can release itself by swimming forward. Remoras sometimes attach to small boats. They swim well on their own, with a sinuous, or curved, motion. When the remora reaches about 3 cm (1.2 in), the disc is fully formed and the remora can then attach to other animals. The remora's lower jaw projects beyond the upper, and the animal lacks a swim bladder. Some remoras associate primarily with specific host species. They are commonly found attached to sharks, manta rays, whales, turtles, and dugongs. Smaller remoras also fasten onto fish such as tuna and swordfish, and some small remoras travel in the mouths or gills of large manta rays, ocean sunfish, swordfish, and sailfish. The remora benefits by using the host as transport and protection, and also feeds on materials dropped by the host.

Some remoras, such as this *Echeneis naucrates*, may attach themselves to scuba divers.

Anglerfish

In some species of anglerfish, when a male finds a female, he bites into her skin, and releases an enzyme that digests the skin of his mouth and her body, fusing the pair down to the blood-vessel level. The male becomes dependent on the female host for survival by receiving nutrients via their shared circulatory system, and provides sperm to the female in return. After fusing, males increase in volume and become much larger relative to free-living males of the species. They live and remain reproductively functional as long as the female lives, and can take part in multiple spawnings. This extreme sexual dimorphism ensures, when the female is ready to spawn, she has a mate immediately available. Multiple males can be incorporated into a single individual female with up to eight males in some species, though some taxa appear to have a one male per female rule.

Parasites

Many endoparasites and ectoparasites, due to their parasitic behaviour, are transported by other animals. For example, tapeworms attach themselves to the inside of the alimentary tracts of other animals and do not locomote within the animal. They do however depend on movement of the host to distribute their eggs.

Other parasites may locomote within, or on, their host, which in turn might be active or stationary. For example, an adult dog flea may crawl about the skin of its sleeping canine host (locomotion), but when the dog awakes and moves, it could be argued the flea is being transported.

Changes between Media

Some animals locomote between different media, e.g., from aquatic to arial. This often requires different modes of locomotion in the different media and may require a distinct transitional locomotor behaviour.

There are a large number of semi-aquatic animals (animals that spend part of their life cycle in water, or generally have part of their anatomy underwater). These represent the major taxons of mammals (e.g., beaver, otter, polar bear), birds (e.g., penguins, ducks), reptiles (e.g., anaconda, bog turtle, marine iguana) and amphibians (e.g., salamanders, frogs, newts).

Fish

Some fish use multiple modes of locomotion. Walking fish may swim freely or at other times "walk" along the ocean or river floor, but not on land (e.g., the flying gurnard —which does not actually fly—and batfishes of the Ogcocephalidae family). Amphibious fish, are fish that are able to leave water for extended periods of time. These fish use a range of terrestrial locomotory modes, such as lateral undulation, tripod-like walking (using paired fins and tail), and jumping. Many of these locomotory modes incorporate multiple combinations of pectoral, pelvic and tail fin movement. Examples include eels, mudskippers and the walking catfish. Flying fish can make powerful, self-propelled leaps out of water into air, where their long, wing-like fins enable gliding flight for considerable distances above the water's surface. This uncommon ability is a natural defense mechanism to evade predators. The flights of flying fish are typically around 50 meters, though they can use updrafts at the leading edge of waves to cover distances of up to 400 m (1,300 ft). They can travel at speeds of more than 70 km/h (43 mph). Maximum altitude is 6 m (20 ft) above the surface of the sea. Some accounts have them landing on ships' decks.

Marine Mammals

Pacific white-sided dolphins porpoising

When swimming, several marine mammals such as dolphins, porpoises and pinnipeds, frequently leap above the water surface whilst maintaining horizontal locomotion. This is done for various reasons. When travelling, jumping can save dolphins and porpoises energy as there is less friction while in the air. This type of travel is known as "porpoising". Other reasons for dolphins and porpoises performing porpoising include orientation, social displays, fighting, non-verbal communication, entertainment and attempting to dislodge parasites. In pinnipeds, two types of porpoising have been identified. "High porpoising" is most often near (within 100 metres) the shore and is often followed by minor course changes; this may help seals get their bearings on beaching or rafting sites. "Low porpoising" is typically observed relatively far (more than 100 metres) from shore and often aborted in favour of anti-predator movements; this may be a way for seals to maximize sub-surface vigilance and thereby reduce their vulnerability to sharks.

Some whales raise their (entire) body vertically out of the water in a behaviour known as "breaching".

Birds

Some semi-aquatic birds use terrestrial locomotion, surface swimming, underwater swimming and flying (e.g., ducks, swans). Diving birds also use diving locomotion (e.g., dippers, aulks). Some birds (e.g., ratites) have lost the primary locomotion of flight. The largest of these, ostriches, when being pursued by a predator, have been known to reach speeds over 70 km/h (43 mph), and can maintain a steady speed of 50 km/h (31 mph), which makes the ostrich the world's fastest two-legged animal: Ostriches can also locomote by swimming. Penguins either waddle on their feet or slide on their bellies across the snow, a movement called *tobogganing*, which conserves energy while moving quickly. They also jump with both feet together if they want to move more quickly or cross steep or rocky terrain. To get onto land, penguins sometimes propel themselves upwards at a great speed to leap out the water.

Changes During the Life-cycle

An animal's mode of locomotion may change considerably during its life-cycle. Barnacles are exclusively marine and tend to live in shallow and tidal waters. They have two nektonic (active swimming) larval stages, but as adults, they are sessile (non-motile) suspension feeders. Frequently, adults are found attached to moving objects such as whales and ships, and are thereby transported (passive locomotion) around the oceans.

Function

Animals locomote for a variety of reasons, such as to find food, a mate, a suitable microhabitat, or to escape predators.

Food Procurement

Animals use locomotion in a wide variety of ways to procure food. Terrestrial methods include ambush predation, social predation, grazing. Aquatic methods include filterfeeding, grazing, ram feeding, suction feeding, protrusion and pivot feeding. Other methods include parasitism and parasitoidism.

Methods of Study

A variety of methods and equipment are used to study animal locomotion:

- Treadmills are used to allow animals to walk or run while remaining stationary with respect to external observers. This technique facilitates filming or recordings of physiological information from the animal (e.g., during studies of energetics). Motorized treadmills are also used to measure the endurance capacity (stamina) of animals.

- Racetracks lined with photocells or filmed while animals run along them are used to measure acceleration and maximal sprint speed.

- Kinematics is the study of the motion of an entire animal or parts of its body. It is typically accomplished by placing visual markers at particular anatomical locations on the animal and then recording video of its movement. The video is often captured from multiple angles, with frame rates exceeding 2000 frames per second when capturing high speed movement. The location of each marker is determined for each video frame, and data from multiple views is integrated to give positions of each point through time. Computers are sometimes used to track the markers, although this task must often be performed manually. The kinematic data can be used to determine fundamental motion attributes such as velocity, acceleration, joint angles, and the sequencing and timing of kinematic events. These fundamental attributes can be used to quantify various higher level attributes, such as the physical abilities of the animal (e.g., its maximum running speed, how steep a slope it can climb), neural control of locomotion, gait, and responses to environmental variation. These, in turn, can aid in formulation of hypotheses about the animal or locomotion in general.

- Force plates are platforms, usually part of a trackway, that can be used to measure the magnitude and direction of forces of an animal's step. When used with kinematics and a sufficiently detailed model of anatomy, inverse dynamics solutions can determine the forces not just at the contact with the ground, but at each joint in the limb.

- Electromyography (EMG) is a method of detecting the electrical activity that occurs when muscles are activated, thus determining which muscles an animal uses for a given movement. This can be accomplished either by surface electrodes (usually in large animals) or implanted electrodes (often wires thinner than a human hair). Furthermore, the intensity of electrical activity can correlate to the level of muscle activity, with greater activity implying (though not definitively showing) greater force.

- Sonomicrometry employs a pair of piezoelectric crystals implanted in a muscle or tendon to continuously measure the length of a muscle or tendon. This is useful because surface kinematics may be inaccurate due to skin movement. Similarly, if an elastic tendon is in series with the muscle, the muscle length may not be accurately reflected by the joint angle.

- Tendon force buckles measure the force produced by a single muscle by measuring the strain of a tendon. After the experiment, the tendon's elastic modulus is determined and used to compute the exact force produced by the muscle. However, this can only be used on muscles with long tendons.

- Particle image velocimetry is used in aquatic and aerial systems to measure the flow of fluid around and past a moving aquatic organism, allowing fluid dynamics calculations to determine pressure gradients, speeds, etc.

- Fluoroscopy allows real-time X-ray video, for precise kinematics of moving bones. Markers opaque to X-rays can allow simultaneous tracking of muscle length.

These methods can be combined. For example, studies frequently combine EMG and kinematics to determine *motor pattern*, the series of electrical and kinematic events that produce a given movement.

Role of Skin in Locomotion

Role of skin in locomotion describes how the integumentary system is involved in locomotion. Typically the integumentary system can be thought of as skin, however the integumentary system also includes the segmented exoskeleton in arthropods and feathers of birds. The primary role of the integumentary system is to provide protection for the body. However, the structure of the skin has evolved to aid animals in their different modes of locomotion. Soft bodied animals such as starfish rely on the arrangement of the fibers in their tube feet for movement. Eels, snakes, and fish use their skin like an external tendon to generate the propulsive forces need for undulatory locomotion. Vertebrates that fly, glide, and parachute also have a characteristic fiber arrangements of their flight membranes that allows for the skin to maintain its structural integrity during the stress and strain experienced during flight.

Soft Bodied Locomotion in Invertebrates

The arms of octopus are muscular hydrostats

The term "Soft Bodied" refers to animals which lack typical systems of skeletal support - included in these are most insect larvae and true worms. Animals that are soft bodied are constrained by the geometry and form of their bodies. However it is the geometry and form of their bodies that generate the forces they need to move. The structure of soft bodied skin can be characterized by a patterned fiber arrangement, which provides the shape and structure for a soft bodied animals. Internal to the patterned fiber layer is typically a liquid filled cavity, which is used to generate hydrostatic pressures for movement. Some animals that exhibit soft bodied locomotion include starfish, octopus, and flatworms.

Hydrostatic Skeleton

A hydrostatic skeleton uses hydrostatic pressure generated from muscle contraction against a liquid filled cavity. The liquid filled cavity is commonly referred to as the hydrostatic body. The liquid within the hydrostatic body acts as an incompressible fluid and the body wall of the hydrostatic body provides a passive elastic antagonist to muscle contraction, which in turn generates a force, which in turn creates movement. This structure plays a role in invertebrate support and locomotor systems and is used for the tube feet in starfish and body of worms . A specialized version of the hydrostatic skeleton is a called a muscular hydrostat, which consists of a tightly packed array of three-dimensional muscle fibers surrounding a hydrostatic body. Examples of muscular hydrostats include the arms of octopus and elephant trunks.

Fiber Arrangement

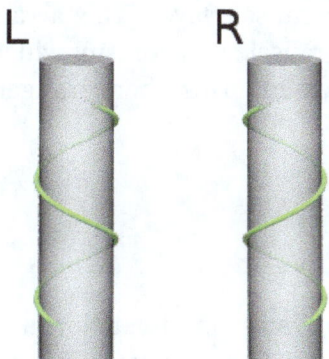

left- and right-handed helices

The arrangement of the connective tissue fibers and muscle fibers create the skeletal support of a soft bodied animal. The arrangement of the fibers around a hydrostatic body limits the range of movement of the hydrostatic body (the "body" of a soft bodied animal) and defines the way the hydrostatic body moves.

Muscle Fibers

Typically muscle fibers surround the hydrostatic body. There are two main types of muscle fibers orientations that are responsible for the movement: the circular orientations and longitudinal orientations. Circular muscles decrease the diameter of a hydrostatic body, resulting in an increase in the length of the body, whereas longitudinal muscles shortens the length of a hydrostatic body, resulting in an increase in the diameter of the body. There are four categories of movements of a hydrostatic skeleton : elongation, shortening, bending and torsion. Elongation, which involves an increase in the length of a hydrostatic body requires either circular muscles, a transverse muscle arrangement, or radial muscle arrangement. For a transverse muscle arrangement, parallel sheets of muscle fibers that extend along the length of a hydrostatic body. For a radial muscle arrangement, radial muscles radiate from a central axis along the axis perpendicular to the long axis. Shortening involves the contraction of the longitudinal muscle. Both shortening and bending involve the contraction of longitudinal muscle, but for bending motion some of the antagonistic muscles work synergistically with longitudinal muscles. The amplitude of movements are based upon the antagonistic muscles forces and the amount of leverage the antagonistic muscle provides for movement. For the torsion motion, muscles are arranged in helical layers around a hydrostatic body. The fiber angle (the angle the fiber makes with the long axis of the body) plays a critical role in torsion, if the angle is greater than 54°44', during muscle contraction, torsion and elongation will occur. If the fiber angle is less than 54°44', torsion and shortening will occur.

Connective tissue Fibers

The arrangement of connective tissue fibers determines the range of motion of a body, and serves as an antagonist against muscle contraction. The most commonly observed connective tissue arrangement for soft bodied animals consists of layers of alternating right and left-handed helices of connective tissue fibers which surround the hydraulic body. This cross helical arrangement is seen in the tube feet starfish, different types of worms and suckers in octopus. This cross helical arrangement allows for the connective tissue layers to evenly distribute force throughout the hydrostatic body. Another commonly observed connective tissue fiber range is when the connective tissue fibers are embedded within a muscle layer. This arrangement of connective tissue fibers creates a stiffer body wall and more muscle antagonism, which allows for more elastic force to be generated and released during movement. This fiber arrangement is seen in the mantle of squid and the fins in sharks.

Specialized Function in Vertebrates

Swimming and Undulatory Locomotion

The skin of these animal that use undulatory motion to locomote have several distinct characteristics. The skin of these animals consists of cross-helical arrangement of collagen and elastin fibers embedded in the dermal layer of skin, a two-dimensional stiffness. which permits bending at small curvatures and resists bending at high curvatures and skin is attached directly to the underlying muscles. Fish, shark, and snakes are all examples of animals that locomote using undulatory locomotion.

Eel

Closeup image of the skin of a marbled eel

The cross helical fiber arrangement of the two dermal fibers types collagen and elastin, are responsible for the mechanical properties of the skin such as the two dimensional stiffness seen in the eel skin. In the longitudinal direction, eel skin behaves like a pure fiber system, with a lessertensile strength than skin in the hoop direction. The skin in the hoop direction exhibits a higher elastic modulus than the skin in the longitudinal direction. The two dimensional stiffness allows for the body of the eel to be modeled a pressurized cylinder with the fiber angle of the cross helical arrangement dictating the method by which the eel moves. Eel skin behaves like skin having a fiber angle greater than 45°. In an eel with the cross helical fiber arrangement, muscle contraction in the anterior region bends the fish, and so the skin on the convex side is extended in the longitudinal direction. The extension in the longitudinal direction produces contraction in the hoop direction as the fiber angle decreases until these dimensional changes are resisted by the body of the eel. The skin becomes skin, and additional longitudinal force(applied by skin) results in force being transmitted along the tail. Therefore, changes in fiber angle of the cross helical arrangement in eel skin allows for the transmission of force through the skin during swimming. The skin act like an external tendon allowing for an eel to generate a greater propulsive force per muscle contraction. In addition to the eel skin acting as an external tendon, the skin attaches directly to the underlying muscle, which allow for the eel to generate an even greater force per muscle contraction.

Longnose Gar

Closeup of the skin of the longnose gar

Due to the heavily scaled skin of the Longnose gar, some of the mechanical properties differ from model of describing how eel skin adds movement. The scale row resists longitudinal forces, which unlike eel skin, makes the skin stiffer in the longitudinal direction, providing myomeres with leverage and anchorage for pulling tendons. At low curvatures, it appears that the dermis is slack on both the concave and convex sides of the body. When the dermis is placed in tension, and resistance to bending is developed, which is referred to as flexural stiffness of the fish skin. The flexural stiffness is a result of the two dimensional stiffness of fish with heavily scaled skin, such as the longnose gar. This mechanical property of fish skin is important to the way a fish swims, because this mechanical property passively stiffens the body, which would otherwise would have been done muscularly. The flexural stiffness of fish skin act in a manner similar to the mechanism by which eel skin acts as an external tendon, however in the case of fish skin, the flexural stiffness acts as a mechanism to decelerate body movement rather than to generate a propulsive force.

Snake

Snakes are one of the few vertebrates in which the skin alone is sufficient for locomotion. During Rectilinear locomotion, the skeleton remains fixed, while the skin is alternately lifted and pulled forward, and then allowed to contact the ground and pulled backwards, propelling the body forward.

One of the interesting aspects of snake skin are folds of intersquamous skin between longitudinally oriented scale rows. The function of these folds is to permit the circumference of the snake to increase, allowing prey to pass into the stomach during feeling. Snakes differ from eels in the direction in which the skin is stiffer, the dorsal scale rows are more flexible in snake than in eels because the dorsal scale row associated with stretching. Differences in the local dermal structures, such as variations in the diameters and orientation of collagen fibers within the intersquamous skin create local differences in the mechanical properties of the snake skin, thus allowing it to adapt to the stresses and strains during the feeding process.

Aerial Locomotion

Gliding, Flying and Parachuting are some of the some methods of aerial locomotion used by animals. Vertebrates have altered the structure of the skin to accommodate the stresses and strains of flight. Typically mammalian skin consists of collagen fibers arranged in a felt-work pattern, with no preferential fiber orientation. However,the structures of skin in bats, birds, and gliding lizards are very different from those of typical mammalian skin. The structural arrangement of the fibers within bat wing skin enables the bat to act like a spring during the down-stroke of flapping. The scales of gliding lizards are arranged in a regular rib like pattern to enable to lizard to act as an airfoil. Avain skin must be structurally arranged such that "the coat of feathers" remains smooth and intact during flight.

Bats

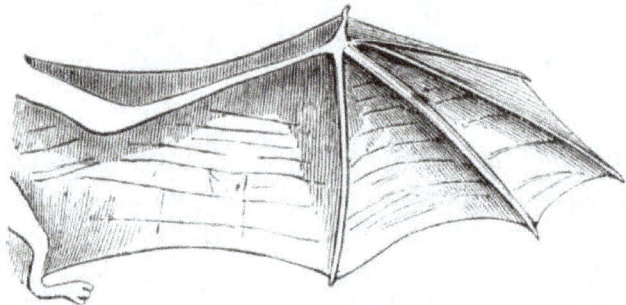

Drawing of the bat wing skin showing the fibers of "mesh like scaffolding"

A closeup view of the felt like fiber pattern seen in elephant skin

Bats rely on skin on their wings to generate lift and thrust used in flight. Therefore, the structure of the bat wing skin is different from the skin of the bat body. Bat wing skin consists of two thin layers of epidermis with a thin layer of dermis/hypodermis located between the epidermal layers whereas the skin of the bat body consists of a single layer of epidermis with a thicker layer of dermis internal to the epidermis. Within the dermal and epidermal layer of bat wing skin, the connective tissue and muscle fibers provide the structural support. The connective tissue fibers within bat wing skin consists of collagen and elastin fiber bundles arranged in a "regular mesh like scaffolding", which the nerves, skeletal muscle fibers and blood vessels embed themselves into. Of the muscles that insert themselves into the mesh scaffolding, larger muscles anchor the skin to the bone and control the membrane tension and camber of the bat wing during flight, whereas smaller muscles, which originate from within the mesh scaffolding, attach to collagen fibers within the fiber network and modulate bone loading and allow for precise control of wing shape and tension. As seen in snakes, local structural differences within the arrangement of the fibers change the mechanical properties of local area, but there are general characteristics that describe the mechanical behavior of bat wing skin. Within the mesh scaffolding of bat wing skin, collagen fibers cross bones perpendicular to the long axes of the bones, therefore mechanical properties of bat wing skin oriented perpendicular to the long axes of the bones exhibit a lower stiffness than the skin that is oriented parallel to the long axes of the bodies. Stiffer skin is necessary for bat wing skin oriented in the direction parallel to the long axes of the bones to prevent too much deformation of bat wing skin during flight(with respect to the bone), resulting in the shearing of the bat wing skin off of the bone. Flexible skin is necessary for the direction perpendicular to the long axes of the bones for facilitating the shape changes needed for movement and control during flight. This anisotropy of bat wing skin is also useful as method of storing and releasing elastic energy, particularly during the downstroke. During the downstroke, the bat extends its wing and the wing skin experiences an aerodynamic force. The wing skin expands and counteracts the aerodynamic force. After the wing is fully extended, the orientation of the wing is nearly vertical and the aerodynamic forces on the wing are reduced. As the aerodynamic force is reduced, the wing recoils, drawing the digits of the wings together in preparation for the upstroke.

Gliding Lizards

There are two different mechanisms by which lizards glide through the air. Both mechanisms involve the patagia.

Active parachuting mechanism

In the active mechanism, skeletal supports and muscles run through the patagia of lizards. The skeletal supports and muscle erect the flight membrane and control the gliding using the patagia. Most of the lizards that exhibit this active gliding mechanism are agamine lizards such lizards in the genus *Draco*. For the passive mechanism of gliding in lizards, the patagia is unfurled by air pressure alone. The patagia of the passive mechanism differs from patagia of the active mechanism; there is the lack of skeleton support and musculature in patagia of the gliding lizards with the passive gliding mechanism. The passive mechanism of gliding is seen in smaller lizards such as the geckos of the genus Ptychozoon. For the passive mechanism of gliding, body movements are believed to control the descent of the gliding lizard.

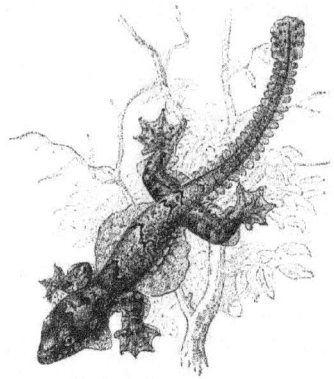

Ptychozoon homalocephalum.

Passive parachuting mechanism

The surface area to body ratios of lizards with different gliding mechanisms are similar, but how the surface area is distributed is different. The difference in the distribution of surface area indicates the differences in the role of the patagia and accessory areas for the different flight mechanisms. Lizards with passive gliding mechanisms tend to have smaller patagia relative to lizards with active flight mechanisms. However,lizards with passive flight mechanism have,ore surface area located in accessory areas(i.e. webbed toes,tail) than lizards with the active gliding mechanism.

The structure of the skin of the patagia and accessory areas for the patagia of a Ptychozoon kuhli, flying gecko, exhibiting the passive gliding mechanism consists of five layers; a layer of adipose tissue is surrounded by a layer of dermis on each side (ventral and dorsal) and a layer of epidermis is external to the two dermal layers. The distribution of the adipose tissue is thickest close to the body wall. This thick layer of adipose tissue at close to the body wall is believed to provide a "safety factor" for the structural elements of the skin (i.e. collagen fibers) near the body wall. The thick layer of adipose tissue is more compliant than the structural elements of the body wall (i.e. ribs, muscles), therefore will more readily deform(absorb force) before the structural elements of the skin experience a force. The layer of adipose tissue also aids in the creation of the domed and cambered shape of the patagia. With regards to the structure of the dermal layer of the patagia, there is a thick layer of collagen fibers oriented parallel to the axis of the patagial ribs. These collagen fibers act as the structural support for the shape of the patagia, and provide the stiffness necessary to resist shape chang . The most prominent features of the epidermal layer of the patagia are the scales. The morphology of the dorsal scales of the patagia change as a result of their functional role. A large portion of the dorsal scales of the patagia are arranged in regular rib-like pattern, which guide the flow of air and allow for the lizard to behave as an airfoil. However at the hinge joints (places where patagia folds and attaches to limbs), the regular rib like structure of scales breakdown into a more random distribution of scales. This breakdown of scales is believed to aid in the mechanical loading of the patagia during the unfurling process and also determining the extent the patagia unfurling during flight.

Birds

The wing of a woodpecker

Avian skin is a bit more complicated than the skin of gliding lizards or bats because the presence of feathers. In addition to the counteracting stresses and stains associated with flight, avian skin must provide a means to monitor and anchor a "coat of feathers", thus the structure of avian skin is different from skin of other flying and gliding animals. To better understand the structure of avian skin, avian skin has been broken down into three different functional components:

- hydraulic skeleto-muscular apparatus of the feathers

This functional component consists of the only of the structural features of the feather tracts, namely the cutis, and connective tissue layer fascia superficialis. This functional component was named "hydraulic skeletal" due to the fact that the fat bodies embedded within cutis and fascia act similar to the hydrostatic bodies within a hydrostatic skeleton. However the functional role of the fat bodies within the hydraulic skeleto-muscular apparatus of the feathers is to counteract forces generated by the erector and depressor muscle of the feathers tracts. rather than to facilitate movement within of a body.

- dermo-subcutaneous muscular system of integument

This functional component of avian skin consists of the smooth muscle of the apertia and striated subcutaneous muscles. The smooth muscles of the apertia counteract the horizontal forces experienced by the feather follicles. The striated subcutaneous muscles also adjust the position of the feather follicles in the directions the smooth muscle cannot. Together this system acts as an integrated muscular system that properly positions the feather tracts on the body of the bird.

- subcutaneous hydraulic skeletal system

This functional component of avian skin consists of the fat bodies of the fascia superficialis and Fascia subcutanea. The majority of the fat bodies are located either between fascia superficialis and the Fascia subcutanea. These fat bodies are stratically located at depression within the body of the bird and function to even out depressions so that feather tracts of the skeleto-muscular apparatus function properly.

Aquatic Locomotion

Aquatic locomotion is biologically propelled motion through a liquid medium. The simplest propulsive systems are composed of cilia and flagella. Swimming has evolved a number of times in a range of organisms including arthropods, fish, molluscs, reptiles, birds, and mammals.

Evolution of Swimming

Swimming evolved a number of times in unrelated lineages. Supposed jellyfish fossils occur in the Ediacaran, but the first free-swimming animals appear in the Early to Middle Cambrian. These are mostly related to the arthropods, and include the Anomalocaridids, which swam by means of lateral lobes in a fashion reminiscent of today's cuttlefish. Cephalopods joined the ranks of the nekton in the late Cambrian, and chordates were probably swimming from the Early Cambrian. Many terrestrial animals retain some capacity to swim, however some have returned to the water and developed the capacities for aquatic locomotion.

Micro-organisms

Ciliates

Ciliates use small flagella called cilia to move through the water. One ciliate will generally have hundreds to thousands of cilia that are densely packed together in arrays. During movement, an individual cilium deforms using a high-friction power stroke followed by a low-friction recovery stroke. Since there are multiple cilia packed together on an individual organism, they display collective behavior in a metachronal rhythm. This means the deformation of one cilium is in phase with the deformation of its neighbor, causing deformation waves that propagate along the surface of the organism. These propagating waves of cilia are what allow the organism to use the cilia in a coordinated manner to move. A typical example of a ciliated microorganism is the *Paramecium*, a one-celled, ciliated protozoan covered by thousands of cilia. The cilia beating together allow the *Paramecium* to propel through the water at speeds of 500 micrometers per second.

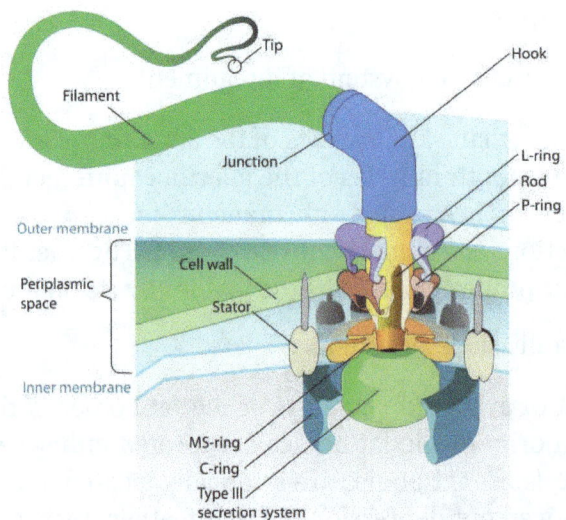

The flagellum of a Gram-negative bacteria is rotated by a molecular motor at its base

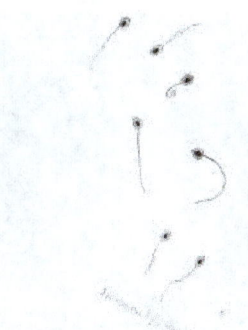

Salmon spermatozoa for artificial propagation

Flagellates

Certain organisms such as bacteria and animal sperm have flagellum which have developed a way to move in liquid environments. A rotary motor model shows that bacteria uses the protons of an electrochemical gradient in order to move their flagella. Torque in the flagella of bacteria is created by particles that conduct protons around the base of the flagellum. The direction of rotation of the flagella in bacteria comes from the occupancy of the proton channels along the perimeter of the flagellar motor.

Movement of sperm is called sperm motility. The middle of the mammalian spermatozoon contains mitochondria that power the movement of the flagellum of the sperm. The motor around the base produces torque, just like in bacteria for movement through the aqueous environment.

Pseudopodia

Movement using a pseudopod is accomplished through increases in pressure at one point on the cell membrane. This pressure increase is the result of actin polymerization between the cortex and the membrane. As the pressure increases the cell membrane is pushed outward creating the pseudopod. When the pseudopod moves outward, the rest of the body is pulled forward by cortical tension. The result is cell movement through the fluid medium. Furthermore, the direction of movement is determined by chemotaxis. When chemoattraction occurs in a particular area of the cell membrane, actin polymerization can begin and move the cell in that direction. An excellent example of an organism that utilizes pseudopods is *Naegleria fowleri*.

Invertebrates

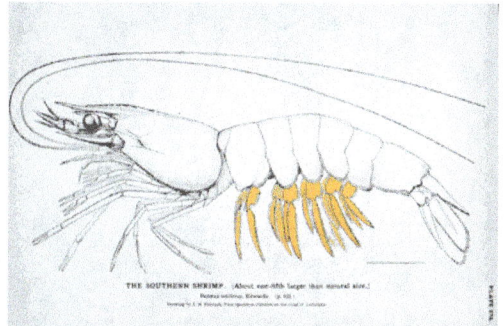

Shrimp paddle with special swimming legs (pleopods)

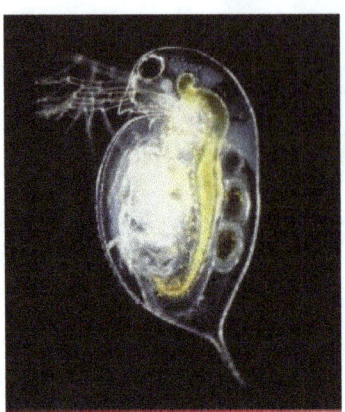

Daphnia swims by beating its antennae

Among the radiata, jellyfish and their kin, the main form of swimming is to flex their cup shaped bodies. All jellyfish are free-swimming, although many of these spend most of their time swimming passively. Passive swimming is akin to gliding; the organism floats, using currents where it can, and does not exert any energy into controlling its position or motion. Active swimming, in contrast, involves the expenditure of energy to travel to a desired location.

In bilateria, there are many methods of swimming. The arrow worms (chaetognatha) undulate their finned bodies, not unlike fish. Nematodes swim by undulating their fin-less bodies. Some Arthropod groups can swim - including many crustaceans. Most crustaceans, such as shrimp, will usually swim by paddling with special swimming legs (pleopods). Swimming crabs swim with modified walking legs (pereiopods). Daphnia, a crustacean, swims by beating its antennae instead.

There are also a number of forms of swimming molluscs. Many free-swimming sea slugs, such as sea angels, flap fin-like structures. Some shelled molluscs, such as scallops can briefly swim by clapping their two shells open and closed. The molluscs most evolved for swimming are the cephalopods.

Among the Deuterostomia, there are a number of swimmers as well. Feather stars can swim by undulating their many arms . Salps move by pumping waters through their gelatinous bodies. The deuterosomes most evolved for swimming are found among the vertebrates, notably the fish.

Jet Propulsion

Octopuses swim headfirst, with arms trailing behind

Jellyfish pulsate their bell for a type of jet locomotion

Jet propulsion is a method of aquatic locomotion where animals fill a muscular cavity and squirt out water to propel them in the opposite direction of the squirting water. Most organisms are equipped with one of two designs for jet propulsion; they can draw water from the rear and expel it from the rear, such as jellyfish, or draw water from front and expel it from the rear, such as salps. Filling up the cavity causes an increase in both the mass and drag of the animal. Because of the expanse of the contracting cavity, the animal's velocity fluctuates as it moves through the water, accelerating while expelling water and decelerating while vacuuming water. Even though these fluctuations in drag and mass can be ignored if the frequency of the jet-propulsion cycles is high enough, jet-propulsion is a relatively inefficient method of aquatic locomotion.

All cephalopods can move by jet propulsion, but this is a very energy-consuming way to travel compared to the tail propulsion used by fish. The relative efficiency of jet propulsion decreases further as animal size increases. Since the Paleozoic, as competition with fish produced an environment where efficient motion was crucial to survival, jet propulsion has taken a back role, with fins and tentacles used to maintain a steady velocity. The stop-start motion provided by the jets, however, continues to be useful for providing bursts of high speed - not least when capturing prey or avoiding predators. Indeed, it makes cephalopods the fastest marine invertebrates, and they can out accelerate most fish. Oxygenated water is taken into the mantle cavity to the gills and through muscular contraction of this cavity, the spent water is expelled through the hyponome, created by a fold in the mantle. Motion of the cephalopods is usually backward as water is forced out anteriorly through the hyponome, but direction can be controlled somewhat by pointing it in different directions. Most cephalopods float (i.e. are neutrally buoyant), so do not need to swim to remain afloat. Squid swim more slowly than fish, but use more power to generate their speed. The loss in efficiency is due to the amount of water the squid can accelerate out of its mantle cavity.

Scallops swim by clapping their two shells open and closed

Jellyfish use a one-way water cavity design which generates a phase of continuous cycles of jet-propulsion followed by a rest phase. The Froude efficiency is about 0.09, which indicates a very costly method of locomotion. The metabolic cost of transport for jellyfish is high when compared to a fish of equal mass.

Other jet-propelled animals have similar problems in efficiency. Scallops, which use a similar design to jellyfish, swim by quickly opening and closing their shells, which draws in water and expels it from all sides. This locomotion is used as a means to escape predators such as starfish. Afterwards, the shell acts as a hydrofoil to counteract the scallop's tendency to sink. The Froude efficiency is low for this type of movement, about 0.3, which is why it's used as an emergency escape mechanism from predators. However, the amount of work the scallop has to do is mitigated by the elastic hinge that connects the two shells of the bivalve. Squids swim by drawing water into their mantle cavity and expelling it through their siphon. The Froude efficiency of their jet-propulsion system is around 0.29, which is much lower than a fish of the same mass.

Much of the work done by scallop muscles to close its shell is stored as elastic energy in abductin tissue, which acts as a spring to open the shell. The elasticity causes the work done against the water to be low because of the large openings the water has to enter and the small openings the water has to leave. The inertial work of scallop jet-propulsion is also low. Because of the low inertial work, the energy savings created by the elastic tissue is so small that it's negligible. Medusae can also use their elastic mesoglea to enlarge their bell. Their mantle contains a layer of muscle sandwiched between elastic fibers. The muscle fibers run around the bell circumferentially while the elastic fibers run through the muscle and along the sides of the bell to prevent lengthening. After making a single contraction, the bell vibrates passively at the resonant frequency to refill the bell. However, in contrast with scallops, the inertial work is similar to the hydrodynamic work due to how medusas expel water - through a large opening at low velocity. Because of this, the negative pressure created by the vibrating cavity is lower than the positive pressure of the jet, meaning that inertial work of the mantle is small. Thus, jet-propulsion is shown as an inefficient swimming technique.

Fish

Open water fish, like this Atlantic bluefin tuna, are usually streamlined for straightline speed, with a deeply forked tail and a smooth body shaped like a spindle tapered at both ends.

Many fish swim through water by creating undulations with their bodies or oscillating their fins. The undulations create components of forward thrust complemented by a rearward force, side forces which are wasted portions of energy, and a normal force that is between the forward thrust and side force. Different fish swim by undulating different parts of their bodies. Eel-shaped fish undulate their entire body in rhythmic sequences. Streamlined fish, such as salmon, undulate the caudal portions of their bodies. Some fish, such as sharks, use stiff, strong fins to create dynamic lift and propel themselves. It is common for fish to use more than one form of propulsion, although they will display one dominant mode of swimming Gait changes have even been observed in juvenile reef fish of various sizes. Depending on their needs, fish can rapidly alternate between synchronized fin beats and alternating fin beats

Many reef fish, like this queen angelfish, have a body flattened like a pancake, with pectoral and pelvic fins that act with the flattened body to maximize manoeuvrability.

According to *Guinness World Records 2009*, *H. zosterae* (the dwarf seahorse) is the slowest moving fish, with a top speed of about 5 feet (150 cm) per hour. They swim very poorly, rapidly fluttering a dorsal fin and using pectoral fins (located behind their eyes) to steer. Seahorses have no caudal fin.

Body-caudal Fin (BCF) Propulsion

- Anguilliform: Anguilliform swimmers are typically slow swimmers. They undulate the majority of their body and use their head as the fulcrum for the load they are moving. At any point during their undulation, their body has an amplitude between 0.5-1.0 wavelengths. The amplitude that they move their body through allows them to swim backwards. Anguilliform locomotion is usually seen in fish with long, slender bodies like eels, lampreys, oarfish, and a number of catfish species.

- Subcarangiform, Carangiform, Thunniform: These swimmers undulate the posterior half of their body and are much faster than anguilliform swimmers. At any point while they are swimming, a wavelength <1 can be seen in the undulation pattern of the body. Some Carangiform swimmers include nurse sharks, bamboo sharks, and reef sharks. Thunniform swimmers are very fast and some common Thunniform swimmers include tuna, white sharks, salmon, jacks, and mako sharks. Thunniform swimmers only undulate their high aspect ratio caudal fin, so they are usually very stiff to push more water out of the way.

- Ostraciiform: Ostraciiform swimmers oscillate their caudal region, making them relatively slow swimmers. Boxfish, torpedo rays, and momyrs employ Ostraciiform locomotion. The cow fish uses Osctraciiform locomotion to hover in the water column.

Median Paired Fin (MPF) Propulsion

- Tetraodoniform, Balistiform, Diodontiform: These swimmers oscillate their median fins. They are typically slow swimmers, and some notable examples include the oceanic sunfish (which has extremely modified anal and dorsal fins), puffer fish, and triggerfish.

- Rajiform, Amiiform, Gymnotiform: This locomotory mode is accomplished by undulation of the pectoral and median fins. During their undulation pattern, a wavelength >1 can be seen in their fins. They are typically slow to moderate swimmers, and some examples include rays, bowfin, and knife fishes. The black ghost knife fish is a Gymnotiform swimmer that has a very long ventral ribbon fin. Thrust is produced by passing waves down the ribbon fin while the body remains rigid. This also allows the ghost knife fish to swim in reverse.

- Labriform: Labriform swimmers are also slow swimmers. They oscillate their pectoral fins to create thrust. Oscillating fins create thrust when a starting vortex is shed from the trailing edge of the fin. As the foil departs from the starting vortex, the effect of that vortex diminishes, while the bound circulation remains, producing lift. Labriform swimming can be viewed as continuously starting and stopping. Wrasses and surf perch are common Labriform swimmers

Hydrofoils

The leopard shark angles its pectoral fins so they behave as hydrofoils to control the animal's pitch

Hydrofoils, or fins, are used to push against the water to create a normal force to provide thrust, propelling the animal through water. Sea turtles and penguins beat their paired hydrofoils to create lift. Some paired fins, such as pectoral fins on leopard sharks, can be angled at varying degrees to allow the animal to rise, fall, or maintain its level in the water column. The reduction of fin surface area helps to minimize drag, and therefore increase efficiency. Regardless of size of the animal, at any particular speed, maximum possible lift is proportional to (wing area) x (speed)^2. Dolphins and whales have large, horizontal caudal hydrofoils, while many fish and sharks have vertical caudal hydrofoils. Porpoising (seen in cetaceans, penguins, and pinnipeds) may save energy if they are moving fast. Since drag increases with speed, the work required to swim unit distance is greater at higher speeds, but the work needed to jump unit distance is independent of speed. Seals propel themselves through the water with their caudal tail, while sea lions create thrust solely with their pectoral flippers.

Drag Powered Swimming

The slowest-moving fishes are the sea horses, often found in reefs

As with moving through any fluid, friction is created when molecules of the fluid collide with organism. The collision causes drag against moving fish, which is why many fish are streamlined in shape. Streamlined shapes work to reduce drag by orienting elongated objects parallel to the force of drag, therefore allowing the current to pass over and taper off the end of the fish. This streamlined shape allows for more efficient use of energy locomotion. Some flat-shaped fish can take advantage of pressure drag by having a flat bottom surface and curved top surface. The pressure drag created allows for the upward lift of the fish.

Appendages of aquatic organisms propel them in two main and biomechanically extreme mechanisms. Some use lift powered swimming, which can be compared to flying as appendages flap like wings, and reduce drag on the surface of the appendage. Others use drag powered swimming, which can be compared to oars rowing a boat, with movement in a horizontal plane, or paddling, with movement in the parasagittal plane.

Drag swimmers use a cyclic motion where they push water back in a power stroke, and return their limb forward in the return or recovery stroke. When they push water directly backwards, this moves their body forward, but as they return their limbs to the starting position, they push water forward, which will thus pull them back to some degree, and so opposes the direction that the body is heading. This opposing force is called drag. The return-stroke drag causes drag swimmers to employ different strategies than lift swimmers. Reducing drag on the return stroke is essential for optimizing efficiency. For example, ducks paddle through the water spreading the webs of their feet as they move water back, and then when they return their feet to the front they pull their webs together to reduce the subsequent pull of water forward. The legs of water beetles have little hairs which spread out to catch up and move water back in the power stroke, but lay flat as the appendage moves forward in the return stroke. Also, the water beetle's legs have a side that is wider and is held perpendicular to the motion when pushing backward, but the leg is then rotated when the limb is to return forward, so that the thinner side will catch up less water.

Drag swimmers experience a lessened efficiency in swimming due to resistance which affects their optimum speed. The less drag a fish experiences, the more it will be able to maintain higher speeds. Morphology of the fish can be designed to reduce drag, such as streamlining the body. The cost of transport is much higher for the drag swimmer, and when deviating from its optimum speed, the drag swimmer is energetically strained much more than the lift swimmer. There are natural processes in place to optimize energy use, and it is thought that adjustments of metabolic rates can compensate in part for mechanical disadvantages.

Semi-aquatic animals compared to fully aquatic animals exhibit exacerbation of drag. Design that allows them to function out of the water limits the efficiency possible to be reached when in the water. In water swimming at the surface exposes them to resistive wave drag and is associated with a higher cost than submerged swimming. Swimming below the surface exposes them to resistance due to return strokes and pressure, but primarily friction. Frictional drag is due to fluid viscosity and morphology characteristics. Pressure drag is due to the difference of water flow around the body and is also affected by body morphology. Semi-aquatic organisms encounter increased resistive forces when in or out of the water, as they are not specialized for either habitat. The morphology of otters and beavers, for example, must meet needs for both environments. Their fur decreases streamlining and creates additional drag. The platypus may be a good example of an intermediate between drag and lift swimmers because it has been shown to have a rowing mechanism which is similar to lift-based pectoral oscillation. The limbs of semi-aquatic organisms are reserved for use on land and using them in water not only increases the cost of locomotion, but limits them to drag-based modes. Although they are less efficient, drag swimmers are able to produce more thrust at low speeds than lift swimmers. They are also thought to be better for maneuverability due to the large thrust produced.

Amphibians

Common toad (*Bufo bufo*) swimming

Most of the Amphibia have a larval state, which has inherited anguilliform motion, and a laterally compressed tail to go with it, from fish ancestors. The corresponding tetrapod adult forms, even in the tail-retaining sub-class Urodeles, are sometimes aquatic to only a negligible extent (as in the genus Salamandra, whose tail has lost its suitability for aquatic propulsion), but the majority of Urodeles, from the newts to the giant salamander Megalobatrachus, retain a laterally compressed tail for a life that is aquatic to a considerable degree, which can use in a carangiform motion.

Of the tailless amphibians (the frogs and toads of the sub-class Anura) the majority are aquatic to an insignificant extent in adult life, but in that considerable minority that are mainly aquatic we encounter for the first time the problem of adapting the tailless-tetrapod structure for aquatic propulsion. The mode that they use is unrelated to any used by fish. With their flexible back legs and webbed feet they execute something close to the leg movements of a human 'breast stroke,' rather more efficiently because the legs are better streamlined.

Reptiles

Nile crocodile (*Crocodylus niloticus*) swimming

From the point of view of aquatic propulsion, the descent of modern members of the class Reptilia from archaic tailed Amphibia is most obvious in the case of the order Crocodilia (crocodiles and alligators), which use their deep, laterally compressed tails in an essentially carangiform mode of propulsion.

Terrestrial snakes, in spite of their 'bad' hydromechanical shape with roughly circular cross-section and gradual posterior taper, swim fairly readily when required, by an anguilliform propulsion

Immature Hawaiian green sea turtle in shallow waters

Cheloniidae (true turtles) have found a beautiful solution to the problem of tetrapod swimming through the development of their forelimbs into flippers of high-aspect-ratio wing shape, with which they imitate a bird's propulsive mode more accurately than do the eagle-rays themselves.

Macroplata

Fin and Flipper Locomotion

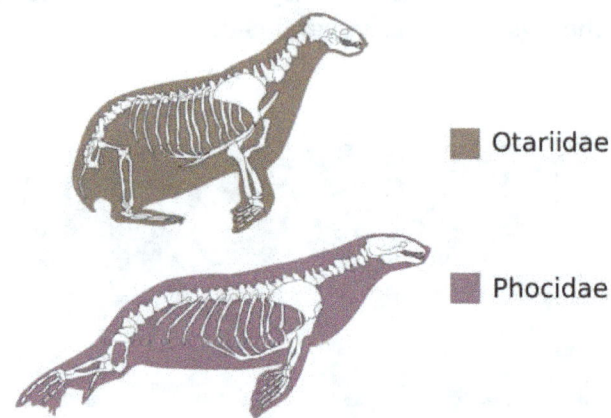

Comparative skeletal anatomy of a typical otariid seal and a typical phocid seal

Aquatic reptiles such as sea turtles and extinct species like Pliosauroids predominantly use their pectoral flippers to propel themselves through the water and their pelvic flippers for maneuvering. During swimming they move their pectoral flippers in a dorso-ventral motion, causing forward motion. During swimming, they rotate their front flippers to decrease drag through the water column and increase efficiency. Newly hatched sea turtles exhibit several behavioral skills that help orientate themselves towards the ocean as well as identifying the transition from sand to water. If rotated in the pitch, yaw or roll direction, the hatchlings are capable of counteracting the forces acting upon them by correcting with either their pectoral or pelvic flippers and redirecting themselves towards the open ocean.

Among mammals otariids (fur seals) swim primarily with their front flippers, using the rear flippers for steering, and phocids (true seals) move the rear flippers laterally, pushing the animal through the water.

Escape Reactions

Some arthropods, such as lobsters and shrimps, can propel themselves backwards quickly by flicking their tail, known as lobstering or the caridoid escape reaction.

Varieties of fish, such as teleosts, also use fast-starts to escape from predators. Fast-starts are characterized by the muscle contraction on one side of the fish twisting the fish into a C-shape. Afterwards, muscle contraction occurs on the opposite side to allow the fish to enter into a steady swimming state with waves of undulation traveling alongside the body. The power of the bending motion comes from fast-twitch muscle fibers located in the central region of the fish. The signal to perform this contraction comes from a set of Mauthner cells which simultaneously send a signal to the muscles on one side of the fish. Mauthner cells are activated when something startles the fish and can be activated by visual or sound-based stimuli.

Fast-starts are split up into three stages. Stage one, which is called the preparatory stroke, is characterized by the initial bending to a C-shape with small delay caused by hydrodynamic resistance. Stage two, the propulsive stroke, involves the body bending rapidly to the other side, which may occur multiple times. Stage three, the rest phase, cause the fish to return to normal steady-state swimming and the body undulations begin to cease. Large muscles located closer to the central portion of the fish are stronger and generate more force than the muscles in the tail. This asymmetry in muscle composition causes body undulations that occur in Stage 3. Once the fast-start is completed, the position of the fish has been shown to have a certain level of unpredictability, which helps fish survive against predators.

The rate at which the body can bend is limited by resistance contained in the inertia of each body part. However, this inertia assists the fish in creating propulsion as a result of the momentum created against the water. The forward propulsion created from C-starts, and steady-state swimming in general, is a result of the body of the fish pushing against the water. Waves of undulation create rearward momentum against the water providing the forward thrust required to push the fish forward.

Efficiency

The Froude propulsion efficiency is defined as the ratio of power output to the power input:

$$nf = 2U_1 / (U_1 + U_2)$$

where U1 = free stream velocity and U2 = jet velocity. A good efficiency for carangiform propulsion is between 50 and 80%.

Minimizing Drag

Pressure differences occur outside the boundary layer of swimming organisms due to disrupted flow around the body. The difference on the up- and down-stream surfaces of the body is pressure drag, which creates a downstream force on the object. Frictional drag, on the other hand, is a result of fluid viscosity in the boundary layer. Higher turbulence causes greater frictional drag.

Reynolds number (Re) is the measure of the relationships between inertial and viscous forces in flow ((animal's length x animal's velocity)/kinematic viscosity of the fluid). Turbulent flow can be found at higher Re values, where the boundary layer separates and creates a wake, and laminar flow can be found at lower Re values, when the boundary layer separation is delayed, reducing wake and kinetic energy loss to opposing water momentum.

The body shape of a swimming organism affects the resulting drag. Long, slender bodies reduce pressure drag by streamlining, while short, round bodies reduce frictional drag; therefore, the optimal shape of an organism depends on its niche. Swimming organisms with a fusiform shape are likely to experience the greatest reduction in both pressure and frictional drag.

Wing shape also affects the amount of drag experienced by an organism, as with different methods of stroke, recovery of the pre-stroke position results in the accumulation of drag.

High-speed ram ventilation creates laminar flow of water from the gills along the body of an organism.

The secretion of mucus along the organism's body surface, or the addition of long-chained polymers to the velocity gradient, can reduce frictional drag experienced by the organism.

Buoyancy

Many aquatic/marine organisms have developed organs to compensate for their weight and control their buoyancy in the water. These structures, make the density of their bodies very close to that of the surrounding water. Some hydrozoans, such as siphonophores, has gas-filled floats; the Nautilus, Sepia, and Spirula (Cephalopods) have chambers of gas within their shells; and most teleost fish and many lantern fish (Myctophidae) are equipped with swim bladders. Many aquatic and marine organisms may also be composed of low-density materials. Deep-water teleosts, which do not have a swim bladder, have few lipids and proteins, deeply ossified bones, and watery tissues that maintain their buoyancy. Some sharks' livers are composed of low-density lipids, such as hydrocarbon squalene or wax esters (also found in Myctophidae without swim bladders), which provide buoyancy.

Swimming animals that are denser than water must generate lift or adapt a benthic lifestyle. Movement of the fish to generate hydrodynamic lift is necessary to prevent sinking. Often, their bodies act as hydrofoils, a task that is more effective in flat-bodied fish. At a small tilt angle, the lift is greater for flat fish than it is for fish with narrow bodies. Narrow-bodied fish use their fins as hydrofoils while their bodies remain horizontal. In sharks, the heterocercal tail shape drives water downward, creating a counteracting upward force while thrusting the shark forward. The lift generated is assisted by the pectoral fins and upward-angle body positioning. It is supposed that tunas primarily use their pectoral fins for lift.

Buoyancy maintenance is metabolically expensive. Growing and sustaining a buoyancy organ, adjusting the composition of biological makeup, and exerting physical strain to stay in motion demands large amounts of energy. It is proposed that lift may be physically generated at a lower energy cost by swimming upward and gliding downward, in a "climb and glide" motion, rather than constant swimming on a plane.

Temperature

Temperature can also greatly affect the ability of aquatic organisms to move through water. This is because temperature not only affects the properties of the water, but also the organisms in the water, as most have an ideal range specific to their body and metabolic needs.

Q10 (temperature coefficient), the factor by which a rate increases at a 10 °C increase in temperature, is used to measure how organisms' performance relies on temperature. Most have increased rates as water becomes warmer, but some have limits to this and others find ways to alter such effects, such as by endothermy or earlier recruitment of faster muscle.

For example, *Crocodylus porosus*, or estuarine crocodiles, were found to increase swimming speed from 15 °C to 23 °C and then to have peak swimming speed from 23 °C to 33 °C. However, performance began to decline as temperature rose beyond that point, showing a limit to the range of temperatures at which this species could ideally perform.

Submergence

The more of the animal's body that is submerged while swimming, the less energy it uses. Swimming on the surface requires two to three times more energy than when completely submerged.

This is because of the bow wave that is formed at the front when the animal is pushing the surface of the water when swimming, creating extra drag.

Secondary Evolution

Chinstrap penguin leaping over water

While tetrapods lost many of their natural adaptations to swimming when they evolved onto the land, many have re-evolved the ability to swim or have indeed returned to a completely aquatic lifestyle.

Primarily or exclusively aquatic animals have re-evolved from terrestrial tetrapods multiple times: examples include amphibians such as newts, reptiles such as crocodiles, sea turtles, ichthyosaurs, plesiosaurs and mosasaurs, marine mammals such as whales, seals and otters, and birds such as penguins. Many species of snakes are also aquatic and live their entire lives in the water. Among invertebrates, a number of insect species have adaptations for aquatic life and locomotion. Examples of aquatic insects include dragonfly larvae, water boatmen, and diving beetles. There are also aquatic spiders, although they tend to prefer other modes of locomotion under water than swimming proper.

Swimming dog

Even though primarily terrestrial tetrapods have lost many of their adaptations to swimming, the ability to swim has been preserved or re-developed in many of them. It may never have been completely lost.

Examples are: Some breeds of dog swim recreationally. Umbra, a world record-holding dog, can swim 4 miles (6.4 km) in 73 minutes, placing her in the top 25% in human long-distance swimming competitions. Although most cats hate water, adult cats are good swimmers. The fishing cat is one wild species of cat that has evolved special adaptations for an aquatic or semi-aquatic lifestyle – webbed digits. Tigers and some individual jaguars are the only big cats known to go into water readily, though other big cats, including lions, have been observed swimming. A few domestic cat breeds also like swimming, such as the Turkish Van.

Horses, moose, and elk are very powerful swimmers, and can travel long distances in the water. Elephants are also capable of swimming, even in deep waters. Eyewitnesses have confirmed that camels, including dromedary and Bactrian camels, can swim, despite the fact that there is little deep water in their natural habitats.

Both domestic and wild rabbits can swim. Domestic rabbits are sometimes trained to swim as a circus attraction. A wild rabbit famously swam in an apparent attack on U.S. President Jimmy Carter's boat when it was threatened in its natural habitat.

The guinea pig (or cavy) is noted as having an excellent swimming ability. Mice can swim quite well. They do panic when placed in water, but many lab mice are used in the Morris water maze, a test to measure learning. When mice swim, they use their tails like flagella and kick with their legs.

Many snakes are excellent swimmers as well. Large adult anacondas spend the majority of their time in the water, and have difficulty moving on land.

Many monkeys can naturally swim and some, like the proboscis monkey, crab-eating macaque, and rhesus macaque swim regularly.

Human Swimming

Swimming has been known amongst humans since prehistoric times; the earliest record of swimming dates back to Stone Age paintings from around 7,000 years ago. Competitive swimming started in Europe around 1800 and was part of the first modern 1896 Summer Olympics in Athens, though not in a form comparable to the contemporary events. It was not until 1908 that regulations were implemented by the International Swimming Federation to produce competitive swimming.

Fish Locomotion

Fish propel themselves through water using many different mechanisms.

Fish locomotion is the variety of types of animal locomotion used by fish, principally by swimming. This however is achieved in different groups of fish by a variety of mechanisms of propulsion in water, most often by wavelike movements of the fish's body and tail, and in various specialised fish by movements of the fins. The major forms of locomotion in fish are anguilliform, in which a wave passes evenly along a long slender body; sub-carangiform, in which the wave increases quickly in amplitude towards the tail; carangiform, in which the wave is concentrated near the tail, which oscillates rapidly; thunniform, rapid swimming with a large powerful crescent-shaped tail; and ostraciiform, with almost no oscillation except of the tail fin. More specialised fish include movement by pectoral fins with a mainly stiff body, as in the sunfish; and movement by propagating a wave along the long fins with a motionless body in fish with electric organs such as the knifefish.

In addition, some fish can variously "walk", i.e., move over land, burrow in mud, and glide through the air.

Swimming

Fish swim by exerting force against the surrounding water. There are exceptions, but this is normally achieved by the fish contracting muscles on either side of its body in order to generate waves of flexion that travel the length of the body from nose to tail, generally getting larger as they go along. The vector forces exerted on the water by such motion cancel out laterally, but generate a net force backwards which in turn pushes the fish forward through the water. Most fishes generate thrust using lateral movements of their body and caudal fin, but many other species move mainly using their median and paired fins. The latter group swim slowly, but can turn rapidly, as is needed when living in coral reefs for example. But they can't swim as fast as fish using their bodies and caudal fins.

Body/Caudal Fin Propulsion

There are five groups that differ in the fraction of their body that is displaced laterally:

Anguilliform

Eels propagate a more or less constant-sized flexion wave along their slender bodies

In the anguilliform group, containing some long, slender fish such as eels, there is little increase in the amplitude of the flexion wave as it passes along the body.

Sub-carangiform

The sub-carangiform group has a more marked increase in wave amplitude along the body with the vast majority of the work being done by the rear half of the fish. In general, the fish body is stiffer, making for higher speed but reduced maneuverability. Trout use sub-carangiform locomotion.

Carangiform

The carangiform group, named for the Carangidae, are stiffer and faster-moving than the previous groups. The vast majority of movement is concentrated in the very rear of the body and tail. Carangiform swimmers generally have rapidly oscillating tails.

Thunniform

The thunniform group contains high-speed long-distance swimmers, and is a unique trait (an autapomorphy) of the tunas. Here, virtually all the sideways movement is in the tail and the region connecting the main body to the tail (the peduncle). The tail itself tends to be large and crescent shaped.

Ostraciiform

The ostraciiform group have no appreciable body wave when they employ caudal locomotion. Only the tail fin itself oscillates (often very rapidly) to create thrust. This group includes Ostraciidae.

Median/Paired Fin Propulsion

Boxfish use median-paired fin swimming, as they are not well streamlined, and use primarily their pectoral fins to produce thrust.

Not all fish fit comfortably in the above groups. Ocean sunfish, for example, have a completely different system, the tetraodontiform mode, and many small fish use their pectoral fins for swimming as well as for steering and dynamic lift. Fish with electric organs, such as those in the knifefish (Gymnotiformes), swim by undulating their very long fins while keeping the body still, presumably so as not to disturb the electric field that they generate.

Many fish swim using combined behavior of their two pectoral fins or both their anal and dorsal fins. Different types of Median paired fin propulsion can be achieved by preferentially using one fin pair over the other, and include rajiform, diodontiform, amiiform, gymnotiform and balistiform modes.

Rajiform

Rajiform locomotion is characteristic of rays, skates, and mantas when thrust is produced by vertical undulations along large, well developed pectoral fins.

Diodontiform

Diodontiform locomotion propels the fish propagating undulations along large pectoral fins, as seen in the porcupinefish (Diodontidae).

Porcupine fish (here, *Diodon nicthemerus*) swim by undulating their pectoral fins.

Amiiform

Amiiform locomotion consists of undulations of a long dorsal fin while the body axis is held straight and stable, as seen in the bowfin.

Gymnotiform

Gymnotus maintains a straight back while swimming to avoid disturbing its electric sense

Gymnotiform locomotion consists of undulations of a long anal fin, essentially upside down amiiform, seen in the knifefish (Gymnotiformes).

Balistiform

In balistiform locomotion, both anal and dorsal fins undulate. It is characteristic of the family Balistidae(triggerfishes). It may also be seen in the Zeidae.

Oscillatory

Oscillation is viewed as pectoral-fin-based swimming and is best known as mobuliform locomotion. The motion can be described as the production of less than half a wave on the fin, similar to a bird wing flapping. Pelagic stingrays, such as the manta, cownose, eagle and bat rays use oscillatory locomotion.

Tetraodontiform

In tetraodontiform locomotion, the dorsal and anal fins are flapped as a unit, either in phase or exactly opposing one another, as seen in the Tetraodontiformes (boxfishes and pufferfishes). The ocean sunfish displays an extreme example of this mode.

Labriform

In labriform locomotion, seen in the wrasses (Labriformes), oscillatory movements of pectoral fins are either drag based or lift based. Propulsion is generated either as a reaction to drag produced by dragging the fins through the water in a rowing motion, or via lift mechanisms.

Dynamic Lift

Sharks are denser than water, and must swim continually, using dynamic lift from their pectoral fins.

Bone and muscle tissues of fish are denser than water. To maintain depth fish such as sharks, but also some bony fish, increase buoyancy by means of a gas bladder or by storing oils or lipids. Fish without these features use dynamic lift instead. It is done using their pectoral fins in a manner similar to the use of wings by airplanes and birds. As these fish swim, their pectoral fins are positioned to create lift which allows the fish to maintain a certain depth. The two major drawbacks of this method are that these fish must stay moving to stay afloat and that they are incapable of swimming backwards or hovering.

Hydrodynamics

Similarly to the aerodynamics of flight, powered swimming requires animals to overcome drag by producing thrust. Unlike flying, however, swimming animals often do not need to supply much vertical force because the effect of buoyancy can counter the downward pull of gravity, allowing these animals to float without much effort. While there is great diversity in fish locomotion, swimming behavior can be classified into two distinct "modes" based on the body structures involved in thrust production, Median-Paired Fin (MPF) and Body-Caudal Fin (BCF). Within each of these classifications, there are numerous specifications along a spectrum of behaviours from purely undulatory to entirely oscillatory. In undulatory swimming modes, thrust is produced by wave-like movements of the propulsive structure (usually a fin or the whole body). Oscillatory modes, on the other hand, are characterized by thrust produced by swiveling of the propulsive structure on an attachment point without any wave-like motion.

Body-caudal Fin

Most fish swim by generating undulatory waves that propagate down the body through the caudal fin. This form of undulatory locomotion is termed Body-Caudal Fin (BCF) swimming on the basis of the body structures used; it includes anguilliform, sub-carangiform, carangiform, and thunniform locomotory modes, as well as the oscillatory ostraciiform mode.

Adaptation

Similar to adaptation in avian flight, swimming behaviors in fish can be thought of as a balance of stability and maneuverability. Because BCF swimming relies on more caudal body structures that can direct powerful thrust only rearwards, this form of locomotion is particularly effective for accelerating quickly and cruising continuously. BCF swimming is, therefore, inherently stable and is often seen in fish with large migration patterns that must maximize efficiency over long periods. Propulsive forces in MPF swimming, on the other hand, are generated by multiple fins located on either side of the body that can be coordinated to execute elaborate turns. As a result, MPF swimming is well adapted for high maneuverability and is often seen in smaller fish that require elaborate escape patterns.

The habitats occupied by fishes are often related to their swimming capabilities. On coral reefs, the faster-swimming fish species typically live in wave-swept habitats subject to fast water flow speeds, while the slower fishes live in sheltered habitats with low levels of water movement.

Fish do not rely exclusively on one locomotor mode, but are rather locomotor generalists, choosing among and combining behaviors from many available behavioral techniques. At slower speeds, predominantly BCF swimmers often incorporate movement of their pectoral, anal, and dorsal fins as an additional stabilizing mechanism at slower speeds, but hold them close to their body at high speeds to improve streamlining and reducing drag. Zebrafish have even been observed to alter their locomotor behavior in response to changing hydrodynamic influences throughout growth and maturation.

In addition to adapting locomotor behavior, controlling buoyancy effects is critical for aquatic survival since aquatic ecosystems vary greatly by depth. Fish generally control their depth by regulating the amount of gas in specialized organs that are much like balloons. By changing the amount of gas in these swim bladders, fish actively control their density. If they increase the amount of air in their swim bladder, their overall density will become less than the surrounding water, and increased upward buoyancy pressures will cause the fish to rise until they reach a depth at which they are again at equilibrium with the surrounding water.

Flight

The transition of predominantly swimming locomotion directly to flight has evolved in a single family of marine fish, the Exocoetidae. Flying fish are not true fliers in the sense that they do not execute powered flight. Instead, these species glide directly over the surface of the ocean water without ever flapping their "wings." Flying fish have evolved abnormally large pectoral fins that act as airfoils and provide lift when the fish launches itself out of the water. Additional forward thrust and steering forces are created by dipping the hypocaudal (i.e. bottom) lobe of their caudal fin into the water and vibrating it very quickly, in contrast to diving birds in which these forces are produced by the same locomotor module used for propulsion. Of the 64 extant species of flying fish, only two distinct body plans exist, each of which optimizes two different behaviors.

Flying fish gain sufficient lift to glide above the water thanks to their enlarged pectoral fins

Tradeoffs

While most fish have caudal fins with evenly sized lobes (i.e. homocaudal), flying fish have an enlarged ventral lobe (i.e. hypocaudal) which facilitates dipping only a portion of the tail back onto the water for additional thrust production and steering.

Because flying fish are primarily aquatic animals, their body density must be close to that of water for buoyancy stability. This primary requirement for swimming, however, means that flying fish are heavier (have a larger mass) than other habitual fliers, resulting in higher wing loading and lift to drag ratios for flying fish compared to a comparably sized bird. Differences in wing area, wing span, wing loading, and aspect ratio have been used to classify flying fish into two distinct classifications based on these different aerodynamic designs.

Biplane Body Plan

In the biplane or *Cypselurus* body plan, both the pectoral and pelvic fins are enlarged to provide lift during flight. These fish also tend to have "flatter" bodies which increase the total lift producing area thus allowing them to "hang" in the air better than more streamlined shapes. As a result of this high lift production, these fish are excellent gliders and are well adapted for maximizing flight distance and duration.

In the monoplane body plan of *Exocoetus*, only the pectoral fins are abnormally large, while the pelvic fins are small.

Comparatively, *Cypselurus* flying fish have lower wing loading and smaller aspect ratios (i.e. broader wings) than their *Exocoetus* monoplane counterparts, which contributes to their ability to fly for longer distances than fish with this alternative body plan. Flying fish with the biplane design take advantage of their high lift production abilities when launching from the water by utilizing a "taxiing glide" in which the hypocaudal lobe remains in the water to generate thrust even after the trunk clears the water's surface and the wings are opened with a small angle of attack for lift generation.

Monoplane Body Plan

In the *Exocoetus* or monoplane body plan, only the pectoral fins are enlarged to provide lift. Fish with this body plan tend to have a more streamlined body, higher aspect ratios (long, narrow wings), and higher wing loading than fish with the biplane body plan, making these fish well adapted for higher flying speeds. Flying fish with a monoplane body plan demonstrate different launching behaviors from their biplane counterparts. Instead of extending their duration of thrust production, monoplane fish launch from the water at high speeds at a large angle of attack (sometimes up to 45 degrees). In this way, monoplane fish are taking advantage of their adaptation for high flight speed, while fish with biplane designs exploit their lift production abilities during takeoff.

Walking

A "walking fish" is a fish that is able to travel over land for extended periods of time. Some other cases of nonstandard fish locomotion include fish "walking" along the sea floor, such as the handfish or frogfish.

Most commonly, walking fish are amphibious fish. Able to spend longer times out of water, these fish may use a number of means of locomotion, including springing, snake-like lateral undulation, and tripod-like walking. The mudskippers are probably the best land-adapted of contemporary fish and are able to spend days moving about out of water and can even climb mangroves, although to only modest heights. The Climbing gourami is often specifically referred to as a "walking fish", although it does not actually "walk", but rather moves in a jerky way by supporting itself on the extended edges of its gill plates and pushing itself by its fins and tail. Some reports indicate that it can also climb trees.

There are a number of fish that are less adept at actual walking, such as the walking catfish. Despite being known for "walking on land", this fish usually wriggles and may use its pectoral fins to aid in its movement. Walking Catfish have a respiratory system that allows them to live out of water for several days. Some are invasive species. A notorious case in the United States is the Northern snakehead. Polypterids have rudimentary lungs and can also move about on land, though rather clumsily. The Mangrove rivulus can survive for months out of water and can move to places like hollow logs.

Ogcocephalus parvus

There are some species of fish that can "walk" along the sea floor but not on land; one such animal is the flying gurnard (it does not actually fly, and should not be confused with flying fish). The batfishes of the Ogcocephalidae family (different than Batfish of Ephippidae) are also capable of walking along the sea floor. Bathypterois grallator, also known as a "tripodfish", stands on its three fins on the bottom of the ocean and hunts for food. The African lungfish (*P. annectens*) can use its fins to *"walk"* along the bottom of its tank in a manner similar to the way amphibians and land vertebrates use their limbs on land.

Burrowing

Many fishes, particularly eel-shaped fishes such as true eels, moray eels, and spiny eels, are capable of burrowing through sand or mud. Ophichthids, the snake eels, are capable of burrowing either forwards or backwards.

Fin and Flipper Locomotion

A species of mudskipper (*Periophthalmus gracilis*)

Fin and flipper locomotion occurs mostly in aquatic locomotion, and rarely in terrestrial locomotion. From the three common states of matter — gas, liquid and solid, these appendages are adapted for liquids, mostly fresh or saltwater and used in locomotion, steering and balancing of the body. Locomotion is important in order to escape predators, acquire food, find mates and bury for shelter, nest or food. Aquatic locomotion consists of swimming, whereas terrestrial locomotion encompasses walking, 'crutching', jumping, digging as well as covering. Some animals such as sea turtles and mudskippers use these two environments for different purposes, for example using the land for nesting, and the sea to hunt for food.

Aquatic Locomotion with Fins and Flippers

Aquatic Locomotion of Fish

Fish live in Fresh or Saltwater habitats and some exceptions are capable of coming on land (Mudskippers). Most fish have a line of muscle blocks, called myomeres, along each side of the body. To swim, they alternately contract one side and relax the other side in a progression which goes from the head to the tail. In this way, an undulatory locomotion results, first bending the body one way in a wave which travels down the body, and then back the other way, with the contracting and relaxing muscles switching roles. They use their fins to propel themselves through the water in this swimming motion. Actinopterygians, the ray-finned fish show an evolutionary pattern of fine control ability to control the dorsal and ventral lobe of the caudal fin. Through developmental changes, intrinsic caudal muscles were added, which enable fish to exhibit such complex maneuvers such as control during acceleration, braking and backing. Studies have shown that the muscles in the caudal fin, have independent activity patterns from the myotomal musculature. These results show specific kinematic roles for different part of the fish's musculature. A curious example of fish adaption is the Ocean sunfish, also known as the *Mola mola*. These fish have undergone significant developmental changes reducing their spinal cord, giving them a disk like appearance, and investing in two very large fins for propulsion. This adaptation usually gives them the appearance that they are as long as they are tall. They are also amazing fish in that they hold the world record in weight gain from fry to adult (60 million times its weight).

Aquatic Locomotion of Marine Mammals

Swimming mammals, such as whales, dolphins,and sea lions, use their flippers to move forward through the water column. During swimming sea lions have a thrust phase, which lasts about 60% of the full cycle, and the recovery phase lasts the remaining 40%. A full cycle duration lasts about 0.5 to 1.0 seconds. Changing direction is a very rapid maneuver that is initiated by head movement towards the back of the animal that is followed by a spiral turn with the body. Due to their pectoral flippers being so closely located to their center of gravity, sea lions are capable of displaying astounding maneuverability in the pitch, roll, and yaw direction and are therefore not constrained, turning stochastically as they please. It is hypothesized that the increased level of maneuverability is caused by their complex habitat. Hunting occurs in difficult environments containing rocky inshore/kelp forest communities, with many niches for prey to hide, therefore requiring speed and maneuverability for capture. The complex skills of a sea lion are learned early on in ontogeny and most are perfected by the time the pups reach one year. Whales and dolphins are less maneuverable and more constrained in their movements. However, dolphins are capable of accelerating as fast as sea lions, but they are not capable of turning as quickly and as efficiently. For both whales and dolphins, their center of gravity does not line up with their pectoral flippers in a straight line, causing a much more rigid and stable swimming pattern.

Aquatic Locomotion of Marine Reptiles

Aquatic reptiles such as sea turtles predominantly use their pectoral flippers to propulse through the water and their pelvic flippers for maneuvering. During swimming they move their pectoral flippers in a clapping motion underneath their body and pull them back up into an airplane position, causing forward motion. During the swimming motion it is really important that they rotate their front flipper in order to decrease drag through the water column and increase their efficiency. Sea turtles exhibit a natural suite of behavior skills that help them direct themselves towards the ocean as well as identify the transition from sand to water after hatching. If rotated in the pitch, yaw or roll direction the hatchlings are capable of counteracting the forces acting upon them by correcting with either their pectoral or pelvic flippers and redirecting themselves towards the open ocean.

Terrestrial Locomotion

Terrestrial Locomotion of Fish

Terrestrial locomotion poses new obstacles such as gravity and new media, including sand, mudd, twigs, logs, debris, grass and many more. Fins and flippers are aquatically adapted appendages and typically aren't very useful in such an environment. It could be hypothesized that fish would try to "swim" on land, but studies have shown that some fish evolved to cope with the terrestrial environment. Mudskippers, for example demonstrate a 'crutching' gait which enables them to 'walk' over muddy surfaces as well as dig burrows to hide in. Mudskippers are also able to jump up to 3 cm distances. This behavior is described as starting with a J-curvature of the body at about 2/3 of its body length (with its tail wrapped towards the head), followed by a straightening of their body which propulses them like a projectile through the air. This behavior enables them to cope with the new environment and opens their habitat to new food sources as well as new predators.

Mudskippers in The Gambia

Terrestrial Locomotion of Marine Reptiles

Caretta caretta Jekyll Island, GA

Reptiles, such as sea turtles spend most of their lives in the ocean. However, their life cycle requires the females to come on shore and lay their nests on the beach. Consequently, the hatchlings emerge from the sand and have to run toward the water. Depending on their species, sea turtles are described to have either a symmetrical gait (diagonally opposite limbs are moving together) or an asymmetrical gait (Contra-lateral limbs move together). For example, loggerhead sea turtle hatchlings are commonly seen exhibiting symmetrical gait on sand, whereas, leatherback sea turtles employ the asymmetrical gait while on land. Notably, leatherbacks employ their front (pelvic) flippers more during forward terrestrial locomotion. Sea turtles can be seen nesting on subtropical and tropical beaches all around the world and exhibit such behavior such as arribada (Collective animal behavior). This is a phenomenon seen in Kemp's Ridley turtles which emerge all at once in one night only onto the beach to lay their nests.

Animal Locomotion on the Water Surface

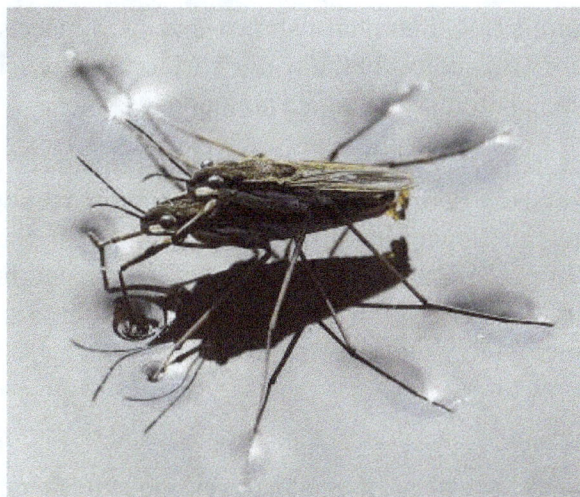

Water striders can move on the surface of water

Animal locomotion on the surface layer of water is the study of animal locomotion in the case of small animals that live on the surface layer of water, relying on surface tension to stay afloat.

There are two types of animal locomotion on water, determined by the ratio of the animal's weight to the water's surface tension: those whose weight is supported by the surface tension at rest, and can therefore easily remain on the water's surface without much exertion, and those whose weight is not supported by the water's surface tension at rest, and must therefore exert additional motion in a direction parallel to the water's surface in order to remain above it. A creature such as the basilisk lizard, often dubbed the 'Jesus lizard', has a weight which is larger than the surface tension can support, and is widely known for running across the surface of water. Another example, the western grebe, performs a mating ritual that includes running across the surface of water.

Surface living animals such as the water strider typically have hydrophobic feet covered in small hairs that prevent the feet from breaking the surface and becoming wet. Another insect known to walk on the water surface is the ant species *Polyrhachis sokolova*. The pygmy gecko (*Coleodactylus amazonicus*), due to its small size and hydrophobic skin is also able to walk on the water surface.

According to biophysicist David L. Hu, there are at least 342 species of water striders. As striders increase in size, their legs become proportionately longer, with *Gigantometra gigas* having a length of over 20 cm requiring a surface tension force of about 40 millinewtons.

Water striders generate thrust by shedding vortices in the water: a series of "U"-shaped vortex filaments is created during the power stroke. The two free ends of the "U" are attached to the water surface. These vortices transfer enough (backward) momentum to the water to propel the animal forwards (some momentum is transferred by capillary waves; Denny's paradox.)

Meniscus Climbing

To pass from the water surface to land, a water-walking insect must contend with the slope of the meniscus at the water's edge. Many such insects are unable to climb this meniscus using their usual propulsion mechanism.

David Hu and coworker John W. M. Bush have shown that such insects climb meniscuses by assuming a fixed body posture. This deforms the water surface and generates capillary forces that propels the insect up the slope without moving its appendages.

Hu and Bush conclude that meniscus climbing is an unusual means of propulsion in that the insect propels itself in a quasi-static configuration, without moving its appendages. Biolocomotion is generally characterized by the transfer of muscular strain energy to the kinetic and gravitational potential energy of the creature, and the kinetic energy of the suspending fluid. In contrast, meniscus climbing has a different energy pathway: by deforming the free surface, the insect converts muscular strain to the surface energy that powers its ascent.

Marangoni Propulsion

Many insects, including some terrestrial insects, can release a surfactant and propel themselves using the Marangoni effect. Hu and Bush report that *Microvelia* can attain a peak speed of 17 cm/s, which is twice its peak walking speed, using Marangoni propulsion.

Marangoni propulsion by a wetting arthropod is precisely analogous to a soap boat but the situation for insects such as water striders is more complex. Hu and Bush state that "for nonwetting arthropods, the transfer of chemical to kinetic energy is more subtle, as the Marangoni stress must be communicated across the creature's complex surface layer".

Sailing

Velella moves by sailing

Velella, the by-the-wind sailor, is a cnidarian with no means of propulsion other than sailing. A small rigid sail projects into the air and catches the wind. *Velella* sails always align along the direction of the wind where the sail may act as an aerofoil, so that the animals tend to sail downwind at a small angle to the wind.

Flying and Gliding Animals

A number of animals have evolved aerial locomotion, either by powered flight or by gliding. Flying and gliding animals (*volant* animals) have evolved separately many times, without any single ancestor. Flight has evolved at least four times, in the insects, pterosaurs, birds, and bats. Gliding has evolved on many more occasions. Usually the development is to aid canopy animals in getting from tree to tree, although there are other possibilities. Gliding, in particular, has evolved among rainforest animals, especially in the rainforests in Asia (most especially Borneo) where the trees are tall and widely spaced. Several species of aquatic animals, and a few amphibians have also evolved to acquire this gliding flight ability, typically as a means of evading predators.

Greylag geese (*Anser anser*). Birds are one of only four taxonomic groups to have evolved powered flight.

Types

Animal aerial locomotion can be divided into two categories—powered and unpowered. In unpowered modes of locomotion, the animal uses on aerodynamics forces exerted on the body due to wind or falling through the air. In powered flight, the animal uses muscular power to generate aerodynamic forces. Animals using unpowered aerial locomotion cannot maintain altitude and speed due to unopposed drag, while animals using powered flight can maintain steady, level flight as long as their muscles are capable of doing so.

Unpowered

These modes of locomotion typically require an animal start from a raised location, converting that potential energy into kinetic energy and using aerodynamic forces to control trajectory and angle of descent. Energy is continually lost to drag without being replaced, thus these methods of locomotion have limited range and duration.

- Falling: decreasing altitude under the force of gravity, using no adaptations to increase drag or provide lift.

- Parachuting: falling at an angle greater than 45° from the horizontal with adaptations to increase drag forces. Very small animals may be carried up by the wind. Some gliding animals may use their gliding membranes for drag rather than lift, to safely descend.

- Gliding flight: falling at an angle less than 45° from the horizontal with lift from adapted aerofoil membranes. This allows slowly falling directed horizontal movement, with stream-lining to decrease drag forces for aerofoil efficiency and often with some maneuverability in air. Gliding animals have a lower aspect ratio (wing length/breadth) than true flyers.

Powered Flight

Powered flight has evolved only four times (in birds, bats, pterosaurs, and insects). It uses muscu-lar power to generate aerodynamic forces and to replace energy lost to drag.

- Flapping: moving wings to produce lift and thrust. May ascend without the aid of the wind, as opposed to gliders and parachuters.

Externally Powered

Ballooning and soaring are not powered by muscle, but rather by external aerodynamic sources of energy: the wind and rising thermals, respectively. Both can continue as long as the source of external power is present. Soaring is typically only seen in species capable of powered flight, as it requires extremely large wings.

- Ballooning: being carried up into the air from the aerodynamic effect on long strands of silk in the wind. Certain silk-producing arthropods, mostly small or young spiders, secrete a special light-weight gossamer silk for ballooning, sometimes traveling great distances at high altitude.

- Soaring: gliding in rising or otherwise moving air that requires specific physiological and morphological adaptations that can sustain the animal aloft without flapping its wings. The rising air is due to thermals, ridge lift or other meteorological features. Under the right conditions, soaring creates a gain of altitude without expending energy. Large wingspans are needed for efficient soaring.

Many species will use multiple of these modes at various times; a hawk will use powered flight to rise, then soar on thermals, then descend via free-fall to catch its prey.

Evolution and Ecology

Gliding and Parachuting

While gliding occurs independently from powered flight, it has some ecological advantages of its own. Gliding is a very energy-efficient way of travelling from tree to tree. An argument made is that many gliding animals eat low energy foods such as leaves and are restricted to gliding because of this, whereas flying animals eat more high energy foods such as fruits, nectar, and insects. In contrast to flight, gliding has evolved independently many times (more than a dozen times among extant vertebrates), however these groups have not radiated nearly as much as have groups of flying animals.

Worldwide, the distribution of gliding animals is uneven as most inhabit rain forests in Southeast Asia. (Despite seemingly suitable rain forest habitats, few gliders are found in India or New Guinea and none in Madagascar.) Additionally, a variety of gliding vertebrates are found in Africa, a family of hylids (flying frogs) lives in South America and several species of gliding squirrels are found in the forests of northern Asia and North America. Various factors produce these disparities. In the forests of Southeast Asia, the dominant canopy trees (usually dipterocarps) are taller than the canopy trees of the other forests. A higher start provides a competitive advantage of further glides and farther travel. Gliding predators may more efficiently search for prey. The lower abundance of insect and small vertebrate prey for carnivorous animals (such as lizards) in Asian forests (In Australia, many mammals (and all mammalian gliders) possess, to some extent, prehensile tails.

An abundance of lianas (woody vines) may obstruct gliders but aid climbers with prehensile tails. The differing populations may relate to the prevalence in South America (compared to Africa or Southeast Asia) of animals with prehensile tails. Arguably, gliding animals prosper in Southeast Asia because the forests are more open (with more room to glide) than those in South America. In dense forests, a prehensile tail enables better tree to tree movement. Also, South American rainforests tend to have more as there are fewer large animals to eat them compared to Africa and Asia lianas.

Because small animals necessarily have higher surface to volume ratios than larger species of similar form, aerodynamic forces have a greater effect on them, resulting in much lower terminal velocity in free fall and amplifying the effects of even small alterations to body surface area. These small changes provide incremental benefits towards further development of gliding.

Powered Flight

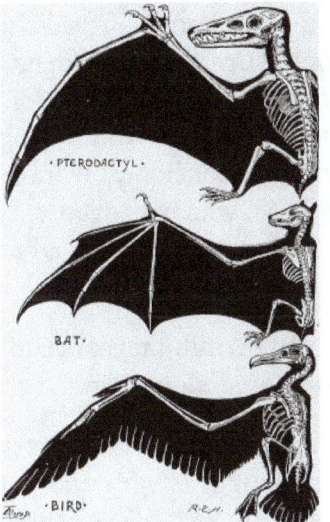

Analogous flying adaptions in vertebrates:
1. pterosaur (Pterosauria)   2. bat (Chiroptera)   3. bird (Aves)

Powered flight has evolved unambiguously only four times—birds, bats, pterosaurs, and insects. In contrast to gliding, which has evolved more frequently but typically gives rise to only a handful of species, all three extant groups of powered flyers have a huge number of species, suggesting that flight is a very successful strategy once evolved. Bats, after rodents, have the most species of any mammalian order, about 20% of all mammalian species. Birds have the most species of any class of terrestrial vertebrates. Finally, insects (most of which fly at some point in their life cycle) have more species than all other animal groups combined.

The evolution of flight is one of the most striking and demanding in animal evolution, and has attracted the attention of many prominent scientists and generated many theories. Additionally, because flying animals tend to be small and have a low mass (both of which increase the surface area to mass ratio), they tend to fossilize infrequently and poorly compared to the larger, heavier-boned terrestrial species they share habitat with. Fossils of flying animals tend to be confined to exceptional fossil deposits formed under highly specific circumstances, resulting in a generally poor fossil record, and a particular lack of transitional forms. Furthermore, as fossils do not preserve behavior or muscle, it can be difficult to discriminate between a poor flyer and a good glider.

Insects were the first to evolve flight, approximately 350 million years ago. The developmental origin of the insect wing remains in dispute, as does the purpose prior to true flight. One suggestion is that wings initially were used to catch the wind for small insects that live on the surface of the water, while another is that they functioned in parachuting, then gliding, then flight for originally arboreal insects.

Pterosaurs were the next to evolve flight, approximately 200 million years ago. These reptiles were close relatives of the dinosaurs (and sometimes mistakenly considered dinosaurs by laymen), and reached enormous sizes, with some of the last forms being the largest flying animals ever to inhabit the Earth, having wingspans of over 9.1 m (30 ft). However, they spanned a large range of sizes, down to a 250 mm (10 in) wingspan in *Nemicolopterus*.

Birds have an extensive fossil record, along with many forms documenting both their evolution from small theropod dinosaurs and the numerous bird-like forms of theropod which did not survive the mass extinction at the end of the Cretaceous. Indeed, *Archaeopteryx* is arguably the most famous transitional fossil in the world, both due to its mix of reptilian and avian anatomy and the luck of being discovered only two years after Darwin's publication of *On the Origin of Species*. However, the ecology and this transition is considerably more contentious, with various scientists supporting either a "trees down" origin (in which an arboreal ancestor evolved gliding, then flight) or a "ground up" origin (in which a fast-running terrestrial ancestor used wings for a speed boost and to help catch prey).

Bats are the most recent to evolve (about 60 million years ago), most likely from a gliding ancestor, though their poor fossil record has hindered more detailed study.

Only a few animals are known to have specialised in soaring: the larger of the extinct pterosaurs, and some large birds. Powered flight is very energetically expensive for large animals, but for soaring their size is an advantage, as it allows them a low wing loading, that is a large wing areas relative to their weight, which maximizes lift. Soaring is very energetically efficient.

Biomechanics

Gliding and Parachuting

During a free-fall with no aerodynamic forces, the object accelerates due to gravity, resulting in increasing velocity as the object descends. During parachuting, animals use the aerodynamic forces on their body to counteract the force or gravity. Any object moving through air experiences a drag force that is proportion to surface area and to velocity squared, and this force will partially counter the force of gravity, slowing the animal's descent to a safer speed. If this drag is oriented at an angle to the vertical, the animal's trajectory will gradually become more horizontal, and it will cover horizontal as well as vertical distance. Smaller adjustments can allow turning or other maneuvers. This can allow a parachuting animal to move from a high location on one tree to a lower location on another tree nearby.

During gliding, lift plays an increased role. Like drag, lift is proportional to velocity squared. Gliding animals will typically leap or drop from high locations such as trees, just as in parachuting, and as gravitational acceleration increases their speed, the aerodynamic forces also increase. Because the animal can utilize lift and drag to generate greater aerodynamic force, it can glide at a shallower angle than parachuting animals, allowing it to cover greater horizontal distance in the same loss of altitude, and reach trees further away.

Unlike most air vehicles, in which the objects that generate lift (wings) and thrust (engine/propeller) are separate and the wings remained fixed, flying animals use their wings to generate both lift and thrust by moving them relative to the body. This has made the flight of organisms considerably harder to understand than that of vehicles, as it involves varying speeds, angles, orientations, areas, and flow patterns over the wings.

A bird or bat flying through the air at a constant speed moves its wings up and down (usually with some fore-aft movement as well). Because the animal is in motion, there is some airflow relative to its body which, combined with the velocity of with wings, generates a faster airflow moving over the wing. This will generate lift force vector pointing forwards and upwards, and a drag force vector pointing rearwards and upwards. The upwards components of these counteract gravity, keeping the body in the air, while the forward component provides thrust to counteract both the drag from the wing and from the body as a whole. Pterosaur flight likely worked in a similar manner, though no living pterosaurs remain for study.

Insect flight is considerably different, due to their small size, rigid wings, and other anatomical differences. Turbulence and vortices play a much larger role in insect flight, making it even more complex and difficult to study than the flight of vertebrates. There are two basic aerodynamic models of insect flight. Most insects use a method that creates a spiralling leading edge vortex. Some very small insects use the fling-and-clap or Weis-Fogh mechanism in which the wings clap together above the insect's body and then fling apart. As they fling open, the air gets sucked in and creates a vortex over each wing. This bound vortex then moves across the wing and, in the clap, acts as the starting vortex for the other wing. Circulation and lift are increased, at the price of wear and tear on the wings.

Limits and Extremes

Flying/Soaring

- Largest. The largest known flying animal was formerly thought to be *Pteranodon*, a pterosaur with a wingspan of up to 7.5 metres (25 ft). However, the more recently discovered azhdarchid pterosaur *Quetzalcoatlus* is much larger, with estimates of the wingspan ranging from 9 to 12 metres (30 to 39 ft). Some other recently discovered azhdarchid pterosaur species, such as *Hatzegopteryx*, may have also wingspans of a similar size or even slightly larger. Although it is widely thought that *Quetzalcoatlus* reached the size limit of a flying animal, it should be noted that the same was once said of *Pteranodon*. The heaviest living flying animals are the kori bustard and the great bustard with males reaching 21 kilograms (46 lb). The wandering albatross has the greatest wingspan of any living flying animal at 3.63 metres (11.9 ft). Among living animals which fly over land, the Andean condor and the marabou stork have the largest wingspan at 3.2 metres (10 ft). Studies have shown that it is physically possible for flying animals to reach 18-metre (59 ft) wingspans, but there is no firm evidence that any flying animal, not even the azhdarchid pterosaurs, got that large.

- Smallest. There is no real minimum size for getting airborne. Indeed, there are many bacteria floating in the atmosphere that constitute part of the aeroplankton. However, to move about under one's own power and not be overly affected by the wind requires a certain amount of size. The smallest flying vertebrates are the bee hummingbird and the bumblebee bat, both of which may weigh less than 2 grams (0.071 oz). They are thought to represent the lower size limit for endotherm flight.

- Fastest. The fastest of all known flying animals is the peregrine falcon, which when diving travels at 300 kilometres per hour (190 mph) or faster. The fastest animal in flapping horizontal flight may be the Mexican free-tailed bat, said to attain about 160 kilometres per hour (99 mph) based on ground speed by an aircraft tracking device; that measurement does not separate any contribution from wind speed, so the observations could be caused by strong tailwinds.

- Slowest. Most flying animals need to travel forward to stay aloft. However, some creatures can stay in the same spot, known as hovering, either by rapidly flapping the wings, as do hummingbirds, hoverflies, dragonflies, and some others, or carefully using thermals, as do some birds of prey. The slowest flying non-hovering bird recorded is the American woodcock, at 8 kilometres per hour (5.0 mph).

- Highest flying. There are records of a Rüppell's vulture *Gyps rueppelli*, a large vulture, being sucked into a jet engine 11,550 metres (37,890 ft) above Côte d'Ivoire in West Africa. The animal that flies highest most regularly is the bar-headed goose *Anser indicus*, which migrates directly over the Himalayas between its nesting grounds in Tibet and its winter quarters in India. They are sometimes seen flying well above the peak of Mount Everest at 8,848 metres (29,029 ft).

Gliding/Parachuting

- Most efficient glider. This can be taken as the animal that moves most horizontal distance per metre fallen. Flying squirrels are known to glide up to 200 metres (660 ft), but have measured glide ratio of about 2. Flying fish have been observed to glide for hundreds of metres on the drafts on the edge of waves with only their initial leap from the water to provide height, but may be obtaining additional lift from wave motion. On the other hand, Albatrosses have measured lift/drag ratios of 20, and thus fall just 1 meter (foot) for every 20 in still air.

- Most maneuverable glider. Many gliding animals have some ability to turn, but which is the most maneuverable is difficult to assess. Even Paradise tree snakes, Chinese gliding frogs, and gliding ants have been observed as having considerable capacity to turn in the air.

Extant Flying and Gliding Animals

Arthropods

- Insects (flying). The first of all animals to evolve flight, insects are also the only invertebrates that have evolved flight. The species are too numerous to list here. Insect flight is an active research field.

  o Gliding bristletails (gliding). Directed aerial gliding descent is found in some tropical arboreal bristletails, an ancestrally wingless sister taxa to the winged insects. The bristletails median caudal filament is important for the glide ratio and gliding control

  o Gliding ants (gliding). The flightless workers of these insects have secondarily gained some capacity to move through the air. Gliding has evolved independently in a number of arboreal ant species from the groups Cephalotini, Pseudomyrmecinae, and Formicinae (mostly *Camponotus*). All arboreal dolichoderines and non-cephalotine myrmicines except *Daceton armigerum* do not glide. Living in the rainforest canopy like many other gliders, gliding ants use their gliding to return to the trunk of the tree they live on should they fall or be knocked off a branch. Gliding was first discovered for *Cephalotes atreus* in the Peruvian rainforest. *Cephalotes atreus* can make 180 degree turns, and locate the trunk using visual cues, succeeding in landing 80% of the time. Unique among gliding animals, Cephalotini and Pseudomyrmecinae ants glide abdomen first, the Forminicae however glide in the more conventional head first manner. The following page has some good videos of gliding ants.

  o Gliding immature insects. The wingless immature stages of some insect species that have wings as adults may also show a capacity to glide. These include some species of cockroach, mantid, katydid, stick insect and true bug.

- Spiders. Although typically flightless some may engage in aerial locomotion as described below.

  o Ballooning spiders (parachuting). The young of some species of spiders travel through the air by using silk draglines to catch the wind, as may some smaller species of adult spider, such the money spider family. This behavior is commonly known as "ballooning". Ballooning spiders make up part of the aeroplankton.

  o Gliding spiders (gliding). Some species of arboreal spider of the genus *Selenops* can glide back to the trunk of a tree should they fall.

Molluscs

- Flying squid (gliding). Several oceanic squids, such as the Pacific flying squid, will leap out of the water to escape predators, an adaptation similar to that of flying fish. Smaller squids will fly in shoals, and have been observed to cover distances as long as 50 metres (160 ft). Small fins towards the back of the mantle do not produce much lift, but do help stabilize the motion of flight. They exit the water by expelling water out of their funnel, indeed some squid have been observed to continue jetting water while airborne providing thrust even after leaving the water. This may make flying squid the only animals with jet-propelled aerial locomotion. The neon flying squid has been observed to glide for distances over 30 metres (100 ft), at speeds of up to 11.2 metres per second (37 ft/s) .

Vertebrates

Fish

Band-winged flying fish, with enlarged pectoral fins

- Flying fish (gliding). There are over 50 species of flying fish belonging to the family Exoco-etidae. They are mostly marine fishes of small to medium size. The largest flying fish can reach lengths of 45 centimetres (18 in) but most species measure less than 30 cm (12 in) in length. They can be divided into two-winged varieties and four-winged varieties. Before the fish leaves the water it increases its speed to around 30 body lengths per second and as it breaks the surface and is freed from the drag of the water it can be traveling at around 60 kilometres per hour (37 mph). The glides are usually up to 30–50 metres (100–160 ft) in length, but some have been observed soaring for hundreds of metres using the updraft on the leading edges of waves. The fish can also make a series of glides, each time dipping the tail into the water to produce forward thrust. The longest recorded series of glides, with the fish only periodically dipping its tail in the water, was for 45 seconds (Video here). It has been suggested that the genus *Exocoetus* is on an evolutionary borderline between flight and gliding. It flaps its enlarged pectoral fins when airborne, but still seems only to glide, as there is no hint of a power stroke. It has been found that some flying fish can glide as effectively as some flying birds.

- Halfbeaks (gliding). A group related to the Exocoetidae, one or two hemirhamphid species possess enlarged pectoral fins and show true gliding flight rather than simple leaps. Marshall (1965) reports that *Euleptorhamphus viridis* can cover 50 metres (160 ft) in two separate hops.

- Freshwater butterflyfish (possibly gliding). *Pantodon buchholzi* has the ability to jump and possibly glide a short distance. It can move through the air several times the length of its body. While it does this, the fish flaps its large pectoral fins, giving it its common name. However, it is debated whether the freshwater butterfly fish can truly glide, Saidel et al. (2004) argue that it cannot.

- Freshwater hatchetfish (possibly flying). There are 9 species of freshwater hatchetfish split among 3 genera. Freshwater hatchetfish have an extremely large sternal region that is fitted with a large amount of muscle that allows it to flap its pectoral fins. They can move in a straight line over a few metres to escape predators.

Illustration of Wallace's flying frog in Alfred Russel Wallace's 1869 book *The Malay Archipelago*

Amphibians

Gliding has evolved independently in two families of tree frogs, the Old World Rhacophoridae and the New World Hylidae. Within each lineage there are a range of gliding abilities from non-gliding, to parachuting, to full gliding.

- Rhacophoridae flying frogs (gliding). A number of the Rhacophoridae, such as Wallace's flying frog (*Rhacophorus nigropalmatus*), have adaptations for gliding, the main feature being enlarged toe membranes. For example, the Malayan flying frog *Rhacophorus prominanus* glides using the membranes between the toes of its limbs, and small membranes located at the heel, the base of the leg, and the forearm. Some of the frogs are quite accomplished gliders, for example, the Chinese flying frog *Rhacophorus dennysi* can maneuver in the air, making two kinds of turn, either rolling into the turn (a banked turn) or yawing into the turn (a crabbed turn).

- Hylidae flying frogs (gliding). The other frog family that contains gliders.

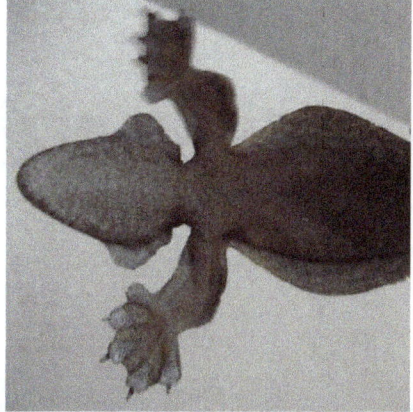

The underside of Kuhl's flying gecko *Ptychozoon kuhli*.The gliding adaptations: flaps of skin on the legs, feet, sides of the body, and on the sides of the head.

Reptiles

Several lizards and snakes are capable of gliding:

- *Draco* lizards. There are 28 species of lizard of the genus *Draco*, found in Sri Lanka, India, and Southeast Asia. They live in trees, feeding on tree ants, but nest on the forest floor. They can glide for up to 60 metres (200 ft) and over this distance they lose only 10 metres (30 ft) in height. Unusually, their patagium (gliding membrane) is supported on elongated ribs rather than the more common situation among gliding vertebrates of having the patagium attached to the limbs. When extended, the ribs form a semicircle on either side the lizard's body and can be folded to the body like a folding fan.

- Gliding lacertids. There are two species of gliding lacertid, of the genus *Holaspis*, found in Africa. They have fringed toes and tail sides and can flatten their bodies for gliding/parachuting.

- *Ptychozoon* flying geckos. There are six species of gliding gecko, of the genus *Ptychozoon*, from Southeast Asia. These lizards have small flaps of skin along their limbs, torso, tail, and head that catch the air and enable them to glide.

- *Lupersaurus* flying geckos. A possible sister-taxon to *Ptychozoon* which has similar flaps and folds and also glides.

- *Thecadactylus* flying geckos. At least some species of *Thecadactylus*, such as *T. rapicauda*, are known to glide.

- *Cosymbotus* flying gecko. Similar adaptations to *Ptychozoon* are found in the two species of the gecko genus *Cosymbotus*.

- *Chrysopelea* snakes. Five species of snake from Southeast Asia, Melanesia, and India. The paradise tree snake of southern Thailand, Malaysia, Borneo, Philippines, and Sulawesi is the most capable glider of those snakes studied. It glides by stretching out its body sideways and opening its ribs so the belly is concave, and by making lateral slithering movements. It can remarkably glide up to 100 metres (330 ft) and make 90 degree turns.

Birds

Birds are a successful group of flying vertebrate.

- Birds (flying, soaring) — Most of the approximately 10,000 living species can fly (flightless birds are the exception). Bird flight is one of the most studied forms of aerial locomotion in animals.

Mammals

Bats are the only mammal with flapping or powered flight. A few other mammals glide or parachute; the best known are flying squirrels and flying lemurs.

- Bats (flying). There are approximately 1,240 bat species, representing about 20% of all classified mammal species.

- Flying squirrels (subfamily Petauristinae) (gliding). There are 43 species divided between 14 genera of flying squirrel. Flying squirrels are found almost worldwide in tropical (Southeast Asia, India, and Sri Lanka), temperate, and even Arctic environments. They tend to be nocturnal. When a flying squirrel wishes to cross to a tree that is further away than the distance possible by jumping, it extends the cartilage spur on its elbow or wrist. This opens out the flap of furry skin (the patagium) that stretches from its wrist to its ankle. It glides spread-eagle and with its tail fluffed out like a parachute, and grips the tree with its claws when it lands. Flying squirrels have been reported to glide over 200 metres (660 ft).

- Anomalures or scaly-tailed flying squirrels (family Anomaluridae) (gliding). These brightly coloured African rodents are not squirrels but have evolved to a resemble flying squirrels by convergent evolution. There are seven species, divided in three genera. All but one species have gliding membranes between their front and hind legs. The genus *Idiurus* contains two particularly small species known as flying mice, but similarly they are not true mice.

- Colugos or "flying lemurs" (order Dermoptera) (gliding). There are two species of colugo. Despite their common name, colugos are not lemurs; true lemurs are primates. Molecular evidence suggests that colugos are a sister group to primates; however, some mammalogists suggest they are a sister group to bats. Found in Southeast Asia, the colugo is probably the mammal most adapted for gliding, with a patagium that is as large as geometrically possible. They can glide as far as 70 metres (230 ft) with minimal loss of height.

- Sifaka, a type of lemur, and possibly some other primates (possible limited gliding/parachuting). A number of primates have been suggested to have adaptations that allow limited gliding and/or parachuting: sifakas, indris, galagos and saki monkeys. Most notably, the sifaka, a type of lemur, has thick hairs on its forearms that have been argued to provide drag, and a small membrane under its arms that has been suggested to provide lift by having aerofoil properties.

- Flying phalangers or wrist-winged gliders (subfamily Petaurinae) (gliding). Possums found in Australia, and New Guinea. The gliding membranes are hardly noticeable until they jump. On jumping, the animal extends all four legs and stretches the loose folds of skin. The subfamily contains seven species. Of the six species in the genus *Petaurus*, the sugar glider and the Biak glider are the most common species. The lone species in the genus *Gymnobelideus*, Leadbeater's possum has only a vestigial gliding membrane.

- Greater glider (*Petauroides volans*) (gliding). The only species of the genus *Petauroides* of the family Pseudocheiridae. This marsupial is found in Australia, and was originally classed with the flying phalangers, but is now recognised as separate. Its flying membrane only extends to the elbow, rather than to the wrist as in Petaurinae.

- Feather-tailed possums (family Acrobatidae) (gliding). This family of marsupials contains two genera, each with one species. The feathertail glider (*Acrobates pygmaeus*), found in Australia is the size of a very small mouse and is the smallest mammalian glider. The feathertail possum (*Distoechurus pennatus*) is found in New Guinea, but does not glide. Both species have a stiff-haired feather-like tail.

Townsends's big-eared bat, (*Corynorhinus townsendii*) displaying the "hand wing"

Extinct Flying and Gliding Animals

Pterosaurs included the largest known flying animals

Reptiles

- Extinct reptiles similar to *Draco* (gliding). There are a number of unrelated extinct lizard-like reptiles with similar "wings" to the *Draco* lizards. *Icarosaurus*, *Coelurosauravus*, *Weigeltisaurus*, *Mecistotrachelos*, and *Kuehneosaurus*. The largest of these, *Kuehneosaurus*, has a wingspan of 30 centimetres (12 in), and was estimated to be able to glide about 30 metres (100 ft).

- Sharovipterygidae (gliding). These strange reptiles from the Upper Triassic of Kyrgyzstan and Poland unusually had a membrane on their elongated hind limbs, extending their otherwise normal, flying-squirrel-like patagia significantly. The forelimbs are in contrast much smaller.

- *Longisquama insignis* (possibly gliding/parachuting). This small reptile may have had long paired feather-like scales on its back, however it has been more recently argued that the scales form just a single dorsal frill. If paired, they may have been used for parachuting. "Everything you can make out is consistent with it being a small, tree-living, gliding animal, which is precisely the thing you'd expect birds to evolve out of," says Larry Martin, senior curator at the Natural History Museum at the University of Kansas.

- Pterosaurs (flying). Pterosaurs were the first flying vertebrates, and are generally agreed to have been sophisticated flyers. They had large wings formed by a patagium stretching from the torso to a dramatically lengthened fourth finger. There were hundreds of species, most of which are thought to have been intermittent flappers, and many soarers. The largest known flying animals are pterosaurs.

- *Hypuronector* (gliding). This bizarre drepanosaur displays limb proportions, particularly the elongated forelimbs, that are consistent with a flying or gliding animal with patagia.

Non-avian Dinosaurs

- Theropods (gliding/flying). There were several species of theropod dinosaur thought to be capable of gliding or flying, that are not classified as birds (though they are closely related). Some species (*Microraptor gui*, *Microraptor zhaoianus*, *Cryptovolans pauli*, and *Changyuraptor*) have been found that were fully feathered on all four limbs, giving them four 'wings' that they are believed to have used for gliding or flying. One species, *Deinonychus antirrhopus*, may display partial volancy, with the young being capable of flight while the adults are flightless, a characteristic also seen in some modern birds like the Horned coot and the Flying steamer duck.

- *Yi* is unique among gliding dinosaurs for the development of membranous wings, unlike the feathered airfoils of other theropods. Much like modern anomalures it developed a bony rod to help support the wing, albeit on the wrist and not the elbow.

Fish

- Thoracopteridae (gliding) is a lineage of Triassic flying fish-like Perleidiformes, having converted their pectoral and pelvic fins into broad wings very similar to those of their modern counterparts. The Ladinian genus *Potanichthys* is the oldest member of this clade, as well as the earliest aerial vertebrate known, suggesting that these fish began exploring aerial niches soon after the Permian-Triassic extinction event.

Mammals

- *Volaticotherium antiquum* (gliding). The earliest known flying or gliding mammal. This squirrel-sized animal belonged to a now extinct ancestral line and was not related to modern day flying or gliding mammals, such as bats or gliding marsupials. It lived around 164 million years ago and used a fur-covered skin membrane to glide through the air. The closely related *Argentoconodon* is also thought to have been able to glide, based on postcranial similarities; it lived around 165 million years ago.

- Several species of extinct bat have been found, like *Icaronycteris*, *Palaeochiropteryx*, and *Onychonycteris*.

- A gliding metatherian (possibly a marsupial) is known from the Paleocene of Itaboraí, Brazil.

Volaticotherids predate bats as mammalian aeronauts by at least 110 million years

Terrestrial Locomotion

Terrestrial locomotion has evolved as animals adapted from aquatic to terrestrial environments. Locomotion on land raises different problems than that in water, with reduced friction being replaced by the effects of gravity.

There are three basic forms of locomotion found among terrestrial animals

- Legged - Moving by using appendages

- Limbless locomotion - moving without legs, primarily using the body itself as a propulsive structure.

- Rolling - rotating the body over the substrate

Legged Locomotion

Movement on appendages is the most common form of terrestrial locomotion, it is the basic form of locomotion of two major groups with many terrestrial members, the vertebrates and the arthropods. Important aspects of legged locomotion are posture (the way the body is supported by the legs), the number of legs, and the functional structure of the leg and foot. There are also many gaits, ways of moving the legs to locomote, such as walking, running, or jumping.

Posture

Appendages can be used for movement in a number of ways. The posture, the way the body is supported by the legs, is an important aspect. There are three main ways in which vertebrates support themselves with their legs - sprawling, semi-erect, and fully erect. Some animals may use different postures in different circumstances, depending on the posture's mechanical advantages. Interestingly, there is no detectable difference in energetic cost between stances.

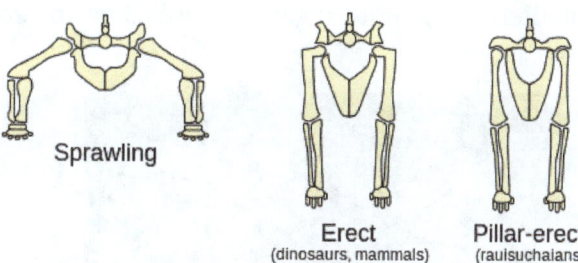

Hip joints and hindlimb postures

The "sprawling" posture is the most primitive, and is the original limb posture from which the others evolved. The upper limbs are typically held horizontally, while the lower limbs are vertical, though upper limb angle may be substantially increased in large animals. The body may drag along the ground, as in salamanders, or may be substantially elevated, as in monitor lizards. This posture is typically associated with trotting gaits, and the body flexes from side-to-side during movement to increase step length. All limbed reptiles and salamanders use this posture, as does the platypus and several species of frogs that walk. Unusual examples can be found among amphibious fish, such as the mudskipper, which drag themselves across land on their sturdy fins. Among the invertebrates, most arthropods—which includes the most diverse group of animals, the insects—have a stance best described as sprawling. There is also anecdotal evidence that some octopus species (such as the *Pinnoctopus* genus) can also drag themselves across land a short distance by hauling their body along by their tentacles (for example to pursue prey between rockpools) - there may be video evidence of this. The semi-erect posture is more accurately interpreted as an extremely elevated sprawling posture. This mode of locomotion is typically found in large lizards such as monitor lizards and tegus.

Mammals and birds typically have a fully erect posture, though each evolved it independently. In these groups the legs are placed beneath the body. This is often linked with the evolution of endothermy, as it avoids Carrier's constraint and thus allows prolonged periods of activity. The fully erect stance is not necessarily the "most-evolved" stance; evidence suggests that crocodilians evolved a semi-erect stance in their forelimbs from ancestors with fully erect stance as a result of adapting to a mostly aquatic lifestyle, though their hindlimbs are still held fully erect. For example, the mesozoic prehistoric crocodilian *Erpetosuchus* is believed to have had a fully erect stance and been terrestrial.

Number of Legs

The number of locomotory appendages varies much between animals, and sometimes the same animal may use different numbers of its legs in different circumstances. The best contender for unipedal movement is the springtail, which while normally hexapedal, hurls itself away from danger using its furcula, a tail-like forked rod that can be rapidly unfurled from the underside of its body.

A number of species move and stand on two legs, that is, they are bipedal. The group that is exclusively bipedal is the birds, which have either an alternating or a hopping gait. There are also a number of bipedal mammals. Most of these move by hopping – including the macropods such as kangaroos and various jumping rodents. Only a few mammals such as humans and the ground pangolin commonly show an alternating bipedal gait. Cockroaches and some lizards may also run on their two hind legs.

The velvet worm (Onychophora)

With the exception of the birds, terrestrial vertebrate groups with legs are mostly quadrupedal – the mammals, reptiles, and the amphibians usually move on four legs. There are many quadrupedal gaits. The most diverse group of animals on earth, the insects, are included in a larger taxon known as hexapods, most of which are hexapedal, walking and standing on six legs. Exceptions among the insects include praying mantises and water scorpions, which are quadrupeds with their front two legs modified for grasping, some butterflies such as the Lycaenidae (blues and hairstreaks) which use only four legs, and some kinds of insect larvae that may have no legs (e.g., maggots), or additional prolegs (e.g., caterpillars).

Spiders and many of their relatives move on eight legs – they are octopedal. However, some creatures move on many more legs. Terrestrial crustaceans may have a fair number - woodlice having fourteen legs. Also, as previously mentioned, some insect larvae such as caterpillars and sawfly larvae have up to five (caterpillars) or nine (sawflies) additional fleshy prolegs in addition to the six legs normal for insects. Some species of invertebrate have even more legs, the unusual velvet worm having stubby legs under the length of its body, with around several dozen pairs of legs. Centipedes have one pair of legs per body segment, with typically around 50 legs, but some species have over 200. The terrestrial animals with the most legs are the millipedes. They have two pairs of legs per body segment, with common species having between 80 and 400 legs overall – with the rare species *Illacme plenipes* having up to 750 legs. Animals with many legs typically move them in metachronal rhythm, which gives the appearance of waves of motion travelling forwards along their rows of legs.

Leg and Foot Structure

The legs of tetrapods, the main group of terrestrial vertebrates, have internal bones, with externally attached muscles for movement, and the basic form has three key joints: the shoulder joint, the knee joint, and the ankle joint, at which the foot is attached. Within this theme there is much variation in form. An alternative form of vertebrate 'leg' to the tetrapod leg is the fins found on amphibious fish. Also a few tetrapods, such as the macropods, have adapted their tails as additional locomotory appendages.

The basic form of the vertebrate foot has five toes, however some animals will have evolved fewer than this, and some early tetrapods had more; Acanthostega had eight toes. Feet have evolved many forms depending on the animal's needs. One key variation is where on the foot the animal's weight is placed. Most vertebrates—the amphibians, the reptiles, and some mammals such as humans and bears—are plantigrade, walking on the whole of the underside of the foot. Many mammals, such as cats and dogs are digitigrade, walking on their toes, the greater stride length allowing more speed. Digitigrade mammals are also often adept at quiet movement. Birds are also digitigrade. Some animals such as horses are unguligrade, walking on the tips of their toes. This even further increases their stride length and thus their speed. A few mammals are also known to walk on their knuckles, at least for their front legs. Knuckle-walking allows the foot (hand) to specialise for food gathering and/or climbing, as with the great apes and the extinct chalicotheres, or for swimming, as with the platypus. In animals where feet have evolved into functional hands, hand walking is also possible.

Among terrestrial invertebrates there are a number of leg forms. The arthropod legs are jointed and supported by hard external armor, with the muscles attached to the internal surface of this exoskeleton. The other group of legged terrestrial invertebrates, the velvet worms, have soft stumpy legs supported by a hydrostatic skeleton. The prolegs that some caterpillars have in addition to their six more-standard arthropod legs have a similar form to those of velvet worms, and suggest a distant shared ancestry.

Gaits

A jumping kangaroo

Animals show a vast range of gaits, the order that they place and lift their appendages in locomotion. Gaits can be grouped into categories according to their patterns of support sequence. For quadrupeds, there are three main categories: walking gaits, running gaits, and leaping gaits. In one system (relating to horses), there are 60 discrete patterns: 37 walking gaits, 14 running gaits, and 9 leaping gaits.

Walking is the most common gait, where some feet are on the ground at any given time, and found in almost all legged animals. In an informal sense, running is considered to occur when at some points in the stride all feet are off the ground in a moment of suspension. Technically, however, moments of suspension occur in both running gaits (such as trot) and leaping gaits (such as canter and gallop). Gaits involving one or more moments of suspension can be found in many animals, and compared to walking they are faster but more energetically costly forms of locomotion.

Animals will use different gaits for different speeds, terrain, and situations. For example, horses show four natural gaits, the slowest horse gait is the walk, then there are three faster gaits which, from slowest to fastest, are the trot, the canter, and the gallop. Animals may also have unusual gaits that are used occasionally, such as for moving sideways or backwards. For example, the main human gaits are bipedal walking and running, but they employ many other gaits occasionally, including a four-legged crawl in tight spaces.

In walking, and for many animals running, the motion of legs on cither side of the body alternates, i.e. is out of phase. Other animals, such as a horse when galloping, or an inchworm, alternate between their front and back legs. An alternative to a gait that alternates between legs is hopping or saltation, where all legs move together. As a main means of locomotion, this is usually found in bipeds or semi-bipeds. Among the mammals saltation is commonly used among macropods (kangaroos and their relatives), jerboas, springhares, kangaroo rats, hopping mice, gerbils, and sportive lemurs. Certain tendons in kangaroo hind legs are very elastic, allowing kangaroos to effectively bounce along conserving energy from hop to hop, making hopping a very energy efficient way to move around in their nutrient poor environment. Saltation is also used by many small birds. Frogs and fleas also hop.

Most animals move in the direction of their head. However, there are some exceptions. Crabs move sideways, and naked mole rats, which live in tight tunnels underground, can move backward or forward with equal facility. Crayfish can move backward much faster than they can move forward.

Gait analysis is the study of gait in humans and other animals. This may involve videoing subjects with markers on particular anatomical landmarks and measuring the forces of their footfall using floor transducers (strain gauges). Skin electrodes may also be used to measure muscle activity.

Limbless Locomotion

A snail moves by slithering

There are a number of terrestrial and amphibious limbless vertebrates and invertebrates. These animals, due to lack of appendages, use their bodies to generate propulsive force. These movements are sometimes referred to as "slithering" or "crawling", although neither are formally used in the scientific literature and the latter term is also used for some animals moving on all four limbs. All limbless animals come from cold-blooded groups; there are no endothermic limbless animals, i.e. there are no limbless birds or mammals.

Lower Body Surface

Where the foot is important to the legged mammal, for limbless animals the underside of the body is important. Some animals such as snakes or legless lizards move on their smooth dry underside. Other animals have various features that aid movement. Molluscs such as slugs and snails move on a layer of mucus that is secreted from their underside, reducing friction and protecting from injury when moving over sharp objects. Earthworms have small bristles (setae) that hook into the substrate and help them move. Some animals such as leeches have suction cups on either end of the body allowing two anchor movement.

Type of Movement

Some limbless animals, such as leeches, have suction cups on either end of their body, which allow them to move by anchoring the rear end and then moving forward the front end, which is then anchored and then the back end is pulled in, and so on. This is known as two-anchor movement. A legged animal, the inchworm, also moves like this, clasping with appendages at either end of its body.

Limbless animals can also move using pedal locomotary waves, rippling the underside of the body. This is the main method used by molluscs such as slugs and snails, and also large flatworms, some other worms, and even earless seals. The waves may move in the opposite direction to motion, known as retrograde waves, or in the same direction as motion, known as direct waves. Earthworms move by retrograde waves alternatively swelling and contracting down the length of their body, the swollen sections being held in place using setae. Aquatic molluscs such as limpets, which are sometimes out of the water, tend to move using retrograde waves. However terrestrial molluscs such as slugs and snails tend to use direct waves. Lugworms and seals also use direct waves.

Most snakes move using lateral undulation where a lateral wave travels down the snake's body in the opposite direction to the snake's motion and pushes the snake off irregularities in the ground. This mode of locomotion requires these irregularities to function. Another form of locomotion, rectilinear locomotion, is used at times by some snakes, especially large ones such as pythons and boa. Here large scales on the underside of the body, known as scutes are used to push backwards and downwards. This is effective on a flat surface and is used for slow, silent movement, such as when stalking prey. Snakes use concertina locomotion for moving slowly in tunnels, here the snake alternates in bracing parts of its body on it surrounds. Finally the caenophidian snakes use the fast and unusual method of movement known as sidewinding on sand or loose soil. The snake cycles through throwing the front part of its body in the direction of motion and bringing the back part of its body into line crosswise.

Rolling

Although animals have never evolved wheels for locomotion, a small number of animals will move at times by rolling their whole body. Rolling animals can be divided into those that roll under the force of gravity or wind and those that roll using their own power.

The pangolin *Manis temminckii* in defensive position

Gravity or Wind Assisted

The web-toed salamander, a 10-centimetre (3.9 in) salamander, lives on steep hills in the Sierra Nevada mountains. When disturbed or startled it coils itself up into a ball, often causing it to roll downhill.

The pebble toad (*Oreophrynella nigra*) lives atop tepui in the Guiana highlands of South America. When threatened, often by tarantulas, it rolls into ball, and typically being on an incline, rolls away under gravity like a loose pebble.

Namib wheeling spiders (*Carparachne spp.*), found in the Namib desert, will actively roll down sand dunes. This action can be used to successfully escape predators such as the *Pompilidae* tarantula wasps, which lay their eggs in a paralyzed spider for their larvae to feed on when they hatch. The spiders flip their body sideways and then cartwheel over their bent legs. The rotation is fast, the golden wheel spider (*Carparachne aureoflava*) moving up to 20 revolutions per second, moving the spider at 1 metre per second.

Coastal tiger beetle larvae when threatened can flick themselves into the air and curl their bodies to form a wheels, which the wind blows, often uphill, as far as 25m and as fast as 11kph. The also may have some ability to steer themselves in this state.

Pangolins, a type of mammal covered in thick scales, roll into a tight ball when threatened. Pangolins have been reported to roll away from danger, by both gravity and self-powered methods. A pangolin in hill country in Sumatra, to flee from the researcher, ran to the edge of a slope and curled into a ball to roll down the slope, crashing through the vegetation, and covering an estimated 30 metres or more in 10 seconds.

Self-powered

Caterpillars of the Mother-Of-Pearl Moth, *Pleuroptya ruralis*, when attacked, will touch their heads to their tails and roll backwards, up to 5 revolutions at about 40 cm per second, which is about 40 times its normal speed.

Nannosquilla decemspinosa, a species of long-bodied, short-legged mantis shrimp, lives in shallow sandy areas along the Pacific coast of Central and South America. When stranded by a low tide the 3 cm stomatopod lies on its back and performs backwards somersaults over and over. The animal moves up to 2 meters at a time by rolling 20–40 times, with speeds of around 72 revolutions per minute. That is 1.5 body lengths per second (3.5 cm/s). Researchers estimate that the stomatopod acts as a true wheel around 40% of the time during this series of rolls. The remaining 60% of the time it has to "jumpstart" a roll by using its body to thrust itself upwards and forwards.

Pangolins have also been reported to roll away from danger by self-powered methods. Witnessed by a lion researcher in the Serengeti in Africa, a group of lions surrounded a pangolin, but could not get purchase on it when it rolled into a ball, and so the lions sat around it waiting and dozing. Surrounded by lions, it would unroll itself slightly and give itself a push to roll some distance, until by doing this multiple times it could get far enough away from the lions to be safe. Moving like this would allow a pangolin to cover distance while still remaining in a protective armoured ball.

Limits and Extremes

The fastest terrestrial animal is the cheetah, which can attain maximal sprint speeds of approximately 104 km/h (64 mph). The fastest running lizard is the Black Iguana, which has been recorded moving at speed of up to 34.9 km/h (21.7 mph).

Arboreal Locomotion

Leopards are good climbers and can carry their kills up their trees to keep them out of reach from scavengers and other predators.

Arboreal locomotion is the locomotion of animals in trees. In habitats in which trees are present, animals have evolved to move in them. Some animals may scale trees only occasionally, but others are exclusively arboreal. The habitats pose numerous mechanical challenges to animals moving through them and lead to a variety of anatomical, behavioral and ecological consequences as well as variations throughout different species. Furthermore, many of these same principles may be applied to climbing without trees, such as on rock piles or mountains.

The earliest known tetrapod with specializations that adapted it for climbing trees was *Suminia*, a synapsid of the late Permian, about 260 million years ago.

Some invertebrate animals are exclusively arboreal in habitat, such as the tree snail.

Biomechanics

Arboreal habitats pose numerous mechanical challenges to animals moving in them, which have been solved in diverse ways. These challenges include moving on narrow branches, moving up and down inclines, balancing, crossing gaps, and dealing with obstructions.

Diameter

Moving along a narrow surface poses special difficulties to animals. During locomotion on the ground, the location of the center of mass may swing from side to side, but during arboreal loco-motion, this would result in the center of mass moving beyond the edge of the branch, resulting in a tendency to topple over. Additionally, foot placement is constrained by the need to make contact with the narrow branch. This narrowness severely restricts the range of movements and postures an animal can use to move.

Incline

Branches are frequently oriented at an angle to gravity in arboreal habitats, including being verti-cal, which poses special problems. As an animal moves up an inclined branch, they must fight the force of gravity to raise their body, making movement more difficult. Conversely, as the animal descends, it must also fight gravity to control its descent and prevent falling. Descent can be partic-ularly problematic for many animals, and highly arboreal species often have specialized methods for controlling their descent.

Balance

Gibbons are very good brachiators because their elongated limbs enable them to easily swing and grasp on to branches.

Due to the height of many branches and the potentially disastrous consequences of a fall, balance is of primary importance to arboreal animals. On horizontal and gently sloped branches, the primary problem is tipping to the side due to the narrow base of support. The narrower the branch, the greater the difficulty in balancing a given animal faces. On steep and vertical branches, tipping becomes less of an issue, and pitching backwards or slipping downwards becomes the most likely failure. In this case, large-diameter branches pose a greater challenge, since the animal cannot place its forelimbs closer to the center of the branch than its hindlimbs.

Crossing Gaps

Branches are not continuous, and any arboreal animal must be able to move between gaps in the branches, or even between trees. This can be accomplished by reaching across gaps, by leaping across them or gliding between them.

Obstructions

Arboreal habitats often contain many obstructions, both in the form of branches emerging from the one being moved on and other branches impinging on the space the animal needs to move through. These obstructions may impede locomotion, or may be used as additional contact points to enhance it. While obstructions tend to impede limbed animals, they benefit snakes by providing anchor points.

Anatomical Specializations

Arboreal organisms display many specializations for dealing with the mechanical challenges of moving through their habitats.

Limb Length

Arboreal animals frequently have elongated limbs that help them cross gaps, reach fruit or other resources, test the firmness of support ahead, and in some cases, to brachiate. However, some species of lizard have reduced limb size that helps them avoid limb movement being obstructed by impinging branches.

Prehensile Tails

Many arboreal species, such as tree porcupines, chameleons, silky anteaters, spider monkeys, and possums, use prehensile tails to grasp branches. In the spider monkey and crested gecko, the tip of the tail has either a bare patch or adhesive pad, which provide increased friction.

Claws

Claws can be used to interact with rough substrates and re-orient the direction of forces the animal applies. This is what allows squirrels to climb tree trunks that are so large as to be essentially flat, from the perspective of such a small animal. However, claws can interfere with an animal's ability to grasp very small branches, as they may wrap too far around and prick the animal's own paw.

The silky anteater uses its prehensile tail as a third arm for stabilization and balance, while its claws help better grasp and climb onto branches.

Adhesion

Adhesion is an alternative to claws, which works best on smooth surfaces. Wet adhesion is common in tree frogs and arboreal salamanders, and functions either by suction or by capillary adhesion. Dry adhesion is best typified by the specialized toes of geckos, which use van der Waals forces to adhere to many substrates, even glass.

Gripping

Frictional gripping is used by primates, relying upon hairless fingertips. Squeezing the branch between the fingertips generates frictional force that holds the animal's hand to the branch. However, this type of grip depends upon the angle of the frictional force, thus upon the diameter of the branch, with larger branches resulting in reduced gripping ability. Animals other than primates that use gripping in climbing include the chameleon, which has mitten-like grasping feet, and many birds that grip branches in perching or moving about.

Reversible Feet

To control descent, especially down large diameter branches, some arboreal animals such as squirrels have evolved highly mobile ankle joints that permit rotating the foot into a 'reversed' posture. This allows the claws to hook into the rough surface of the bark, opposing the force of gravity.

Low Center of Mass

Many arboreal species lower their center of mass to reduce pitching and toppling movement when climbing. This may be accomplished by postural changes, altered body proportions, or smaller size.

Small Size

Small size provides many advantages to arboreal species: such as increasing the relative size of branches to the animal, lower center of mass, increased stability, lower mass (allowing movement on smaller branches), and the ability to move through more cluttered habitat. Size relating to weight affects gliding animals such as the reduced weight per snout-vent length for 'flying' frogs.

Hanging Under Perches

The gecko's toes adhere to surfaces via dry adhesion, to allow them to stay firmly attached to a branch or even a flat wall.

Some species of primate, bat, and all species of sloth achieve passive stability by hanging beneath the branch. Both pitching and tipping become irrelevant, as the only method of failure would be losing their grip.

Behavioral Specializations

Arboreal species have behaviors specialized for moving in their habitats, most prominently in terms of posture and gait. Specifically, arboreal mammals take longer steps, extend their limbs further forwards and backwards during a step, adopt a more 'crouched' posture to lower their center of mass, and use a diagonal sequence gait.

Ecological Consequences

Arboreal locomotion allows animals access to different resources, depending upon their abilities. Larger species may be restricted to larger-diameter branches that can support their weight, while smaller species may avoid competition by moving in the narrower branches.

Climbing Without Trees

Many animals climb in other habitats, such as in rock piles or mountains, and in those habitats, many of the same principles apply due to inclines, narrow ledges, and balance issues. However, less research has been conducted on the specific demands of locomotion in these habitats.

Perhaps the most exceptional of the animals that move on steep or even near vertical rock faces by careful balancing and leaping are the various types of mountain dwelling caprid such as the Barbary sheep, markhor, yak, ibex, tahr, rocky mountain goat, and chamois. Their adaptations may include a soft rubbery pad between their hooves for grip, hooves with sharp keratin rims for lodging in small footholds, and prominent dew claws. The snow leopard, being a predator of such mountain caprids, also has spectacular balance and leaping abilities; being able to leap up to ≈17m (~50 ft). Other balancers and leapers include the mountain zebra, mountain tapir, and hyraxes.

Brachiation

Arboreal snails use their sticky slime to help in climbing up trees, since they lack limbs to do so.

Brachiation is a specialized form of arboreal locomotion, used by primates to move very rapidly while hanging beneath branches. Arguably the epitome of arboreal locomotion, it involves swinging with the arms from one handhold to another. Only a few species are brachiators, and all of these are primates; it is a major means of locomotion among spider monkeys and gibbons, and is occasionally used by female orangutans. Gibbons are the experts of this mode of locomotion, swinging from branch to branch distances of up to 15 m (50 ft), and traveling at speeds of as much as 56 km/h (35 mph).

Gliding and Parachuting

To bridge gaps between trees, many animals such as the flying squirrel have adapted membranes, such as patagia for gliding flight. Some animals can slow their descent in the air using a method known as parachuting, such as *Rhacophorus* (a "flying frog" species) that has adapted toe membranes allowing it to fall more slowly after leaping from trees.

Limbless Climbing

Many species of snake are highly arboreal, and some have evolved specialized musculature for this habitat. While moving in arboreal habitats, snakes move slowly along bare branches using a specialized form of concertina locomotion, but when secondary branches emerge from the branch being moved on, snakes use lateral undulation, a much faster mode. As a result, snakes perform best on small perches in cluttered environments, while limbed organisms seem to do best on large perches in uncluttered environments.

Arboreal Animals

Many species of animals are arboreal, far too many to list individually. This list is of prominently or predominantly arboreal species and higher taxa.

- Primates
- Cats
- brushtail possums
- opossums
- Sloths
- Anteaters
- Treeshrews
- Goats
- Colugos
- Kinkajous
- Viverrids
- Tree squirrels and many other rodents
- Parrots
- Geckos
- Chameleons
- Many other lizards
- Mambas
- Brown Tree Snakes
- Many other snakes
- Stick insects
- Many other arthropods
- Tree snails
- Koalas

References

- Weis, J.S. (2012). Walking sideways: the remarkable world of crabs. Ithaca, NY: Cornell University Press. pp. 63–77. ISBN 978-0-8014-5050-1. OCLC 794640315

- Dewar, H.; Graham, J. (1994). "Studies of tropical tuna swimming performance in a large water tunnel-kinematics". Journal of Experimental Biology. 192 (1): 45–59

- "Spanner crab Ranina ranina". Fishing and Aquaculture. New South Wales Department of Primary Industries. 2005. Retrieved January 4, 2009

- Walker, J.A. and Westneat, M.W. (2000). "Mechanical performance of aquatic rowing and flying". Proceedings of the Royal Society of London B: Biological Sciences. 267 (1455): 1875–1881. doi:10.1098/rspb.2000.1224. CS1 maint: Multiple names: authors list (link)

- Campbell, Neil A.; Reece, Jane B. (2005). Biology, 7th Edition. San Francisco: Pearson - Benjamin Cummings. pp. 522–523. ISBN 0-8053-7171-0

- Sleinis, S.; Silvey, G.E. (1980). "Locomotion in a forward walking crab". Journal of Comparative Physiology A. 136 (4): 301–312. doi:10.1007/BF00657350

- Schroder, G.D. (August 1979). "Foraging behavior and home range utilization of the Bannertail Kangaroo Rat". Ecology. Ecological Society of America. 60 (4): 657–665. JSTOR 1936601. doi:10.2307/1936601

- King, R.S. (2013). "BiLBIQ: A Biologically Inspired Robot with Walking and Rolling Locomotion". Biosystems and Biorobotics. 2. Springer, Verlag, Berlin, Heidelberg. ISBN 978-3-642-34681-1. doi:10.1007/978-3-642-34682-8

- Quillan, K.J. (2000). "Ontogenetic scaling of burrowing forces in the earthworm Lumbricus terrestris" (PDF). Journal of Experimental Biology. 203 (Pt 18): 2757–2770. PMID 10952876

- "Merriam's Kangaroo Rat Dipodomys merriami". U. S. Bureau of Land Management web site. Bureau of Land Management. Retrieved 2014-03-26

- Bowerman, R.F. (1975). "The control of walking in the scorpion". Journal of Comparative Physiology. 100 (3): 183–196. doi:10.1007/bf00614529

- Jurmain, Robert; Kilgore, Lynn; Trevathan, Wenda (2008). Essentials of Physical Anthropology (7 ed.). Cengage Learning. p. 109. ISBN 9780495509394

- Weyman, G.S. (1995). "Laboratory studies of the factors stimulating ballooning behavior by Linyphiid spiders (Araneae, Linyphiidae)" (PDF). The Journal of Arachnology. 23: 75–84. Retrieved 2009-07-18

- Prostak, S. (May 6, 2014). "Cebrennus rechenbergi: Cartwheeling spider discovered in Morocco". Sci-News.com. Retrieved October 20, 2016

- Bejan, Adrian; Marden, James H. (2006). "Constructing Animal Locomotion from New Thermodynamics Theory". American Scientist. 94 (4): 342–349. doi:10.1511/2006.60.342

- Wilbur, Karl M.; Clarke, M.R.; Trueman, E.R., eds. (1985), The Mollusca, 12. Paleontology and neontology of Cephalopods, New York: Academic Press, ISBN 0-12-728702-7

- S. M. Swartz, M. S. Groves, H. D. Kim, and W. R. Walsh. Mechanical properties of bat wing membrane skin. Journal of Zoology,239(2):357–378, 1996

- Roberts, Tristan David Martin (1995). Understanding Balance: The Mechanics of Posture and Locomotion. Nelson Thornes. p. 211. ISBN 978-1-56593-416-0. Retrieved 18 March 2015

- "Sharksucker fish's strange disc explained". Natural History Museum. 28 January 2013. Archived from the original on 1 February 2013. Retrieved 5 February 2013

- Weihs, D. (2002). "Dynamics of Dolphin Porpoising Revisited". Integrative and Comparative Biology. 42 (5): 1071–1078. PMID 21680390. doi:10.1093/icb/42.5.1071

Permissions

Index